Botanical Treasures

Objects from the Herbarium and Library
of the Royal Botanic Garden Edinburgh

Contributing Authors

Hannah Atkins – Deputy Herbarium Curator
Helen Bennett – Library Volunteer
Alan Elliott – PhD Student
Martin Gardner – ICCP Programme Coordinator
Erzsébet Gyöngy – Herbarium Assistant
Graham Hardy – Serials Librarian
David Harris – Herbarium Curator
Elspeth Haston – Deputy Herbarium Curator
Stephan Helfer – Senior Mycologist
Mark Hughes – Tropical Botanist
Sabina Knees – Research Scientist
David Long – Research Associate
Heather McHaffie – Research Associate
Lorna Mitchell – Head of Library Services
Henry Noltie – Researcher, Historic Collections and Archives
Leonie Paterson – Archivist
James Richardson – Tropical Botanist
Adele Smith – Assistant Herbarium Curator
Mark Watson – Head of Major Floras
Rebecca Yahr – Lichen Biodiversity Scientist

Photography

Amy Fokinther

ISBN: 978-1-906129-97-2

Published by
Royal Botanic Garden Edinburgh
20A Inverleith Row, Edinburgh, EH3 5LR
www.rbge.org.uk

Proceeds from sales of this book will be used to support the work of RBGE.
The Royal Botanic Garden Edinburgh is a Non Departmental Public Body (NDPB) sponsored and supported
through Grant-in-Aid by the Scottish Government's Environment and Forestry Directorate (ENFOR).
The Royal Botanic Garden Edinburgh is a Charity registered in Scotland (number SC007983).

The correct citation for this publication is: Royal Botanic Garden Edinburgh. (2014). *Botanical Treasures: Objects from the Herbarium and Library of the Royal Botanic Garden Edinburgh*. Royal Botanic Garden Edinburgh.

Front cover: see p. 29. Back cover: (top) see p. 83; (middle) see p. 139; (bottom) see p. 63.

Printed by Meigle Colour Printers Limited, Galashiels
using vegetable-based inks and eco-friendly varnish,
under the control of an environmental Management System.

Contents

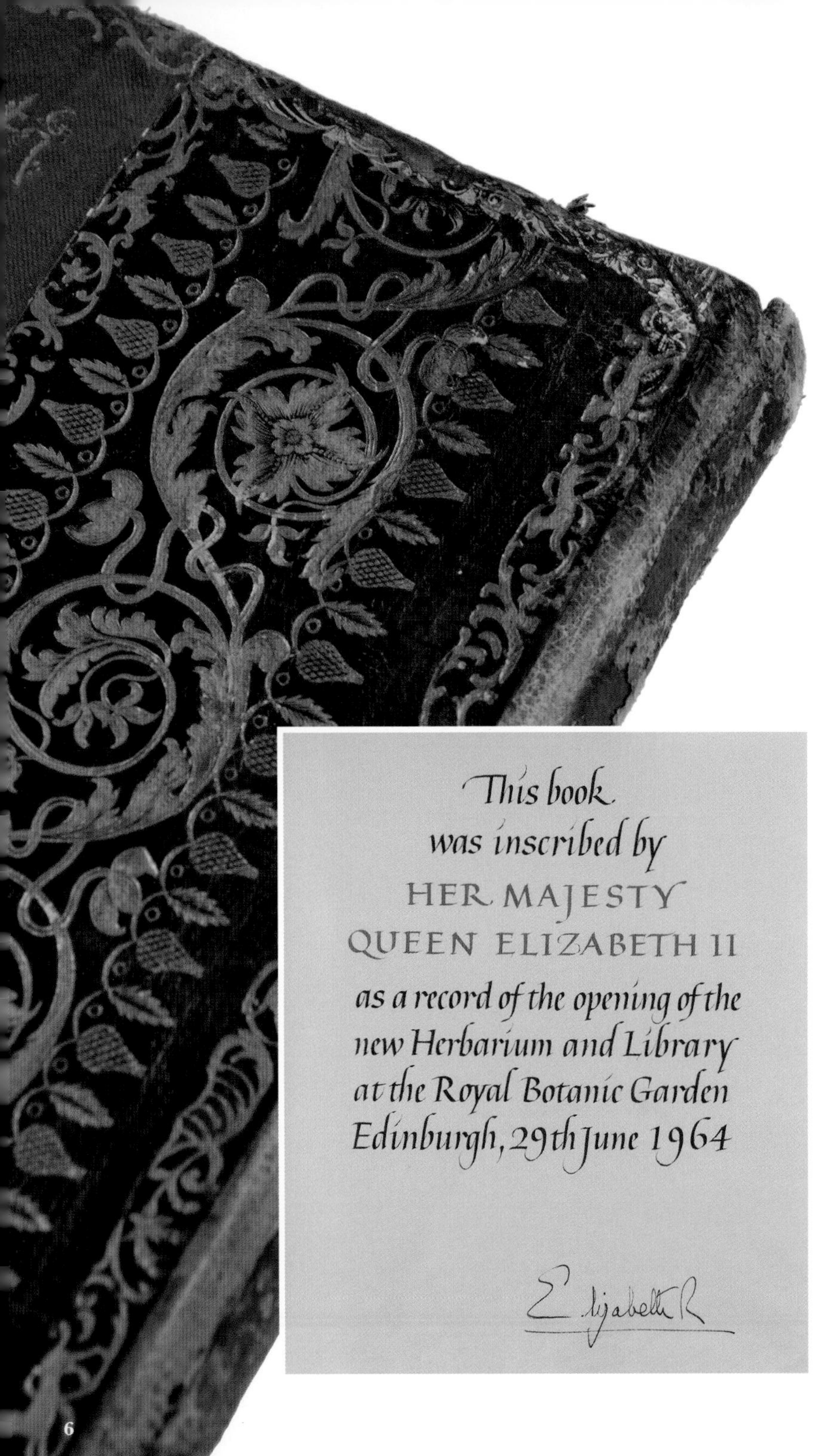

This book
was inscribed by
HER MAJESTY
QUEEN ELIZABETH II
as a record of the opening of the
new Herbarium and Library
at the Royal Botanic Garden
Edinburgh, 29th June 1964

Elizabeth R

A page from *Flowers Drawn and Painted After Nature in India*, c.1835, by Mrs James Cookson, inscribed by Her Majesty The Queen on the occasion of the opening of the new RBGE Herbarium and Library building on 29 June 1964.

One of the hand-coloured lithographic plates from the book, of *Gloriosa superba*, is below.

Regius Keeper's Foreword

It is both a great honour and very humbling to write the foreword for this splendid book so early in my tenure as Regius Keeper. Thanks to the commitment and endeavours of staff and donors over the past 344 years, the Royal Botanic Garden Edinburgh has an extraordinary inheritance: its living, preserved and literary collections; as well as art and artefacts. I remember the sense of amazement when I was first introduced to the collections, and am delighted that this book provides a similar opportunity for a wider audience to have an insight into some of our 'treasures'. Indeed, much of the content is being published for the first time.

Whilst some of the collections are hidden from view within the walls of the Herbarium, Library and Archive, it is important to note that they are fundamental to our research and education activities. The collections underpin our vital contribution to halting the loss of our biodiversity – the staggering diversity of plants and fungi upon which our future health, prosperity and well-being depend.

This book also helps us to appreciate the intrinsic value of the world of plants, reminding us of the beauty of the natural world and the expertise of mankind in capturing and preserving that beauty. It is my hope that this book will enable even more people to be inspired to value our national collections and to support the work of the Royal Botanic Garden Edinburgh.

Simon Milne MBE
Regius Keeper
Royal Botanic Garden Edinburgh

307 MALVACEAE
52 Tilia
65 Helicteres
307 MALVACEAE

Introduction

The Royal Botanic Garden Edinburgh (RBGE) is one of Scotland's most popular attractions with more than 700,000 visitors coming to our Edinburgh Garden each year. But as they stroll around the Garden enjoying the Living Collection, the vast majority of these visitors are probably unaware of the treasures that lie behind the scenes in the large white building which is home to our other collections. This book celebrates the 50th anniversary of the Herbarium and Library building, opened by Her Majesty Queen Elizabeth II on 29 June 1964, by providing a glimpse of some of the treasures in the collections it was built to house.

A herbarium is a collection of dried specimens arranged in a systematic order, serving as a reference library for plant and fungal life. The basic concept of a herbarium specimen is very simple, and the technique used to create them now is the same as was used for our earliest specimen, collected in 1697 (p. 17). A small plant, or part of a larger plant, is pressed flat, dried and mounted on a piece of archival card known as a herbarium sheet. This is stored along with a label containing information on where the plant was collected, when and by whom. When a new species is discovered the name is associated with a specific herbarium specimen (the "type specimen") that defines the application of the name.

Above: The RBGE Herbarium and Library building.

Below: A herbarium specimen.

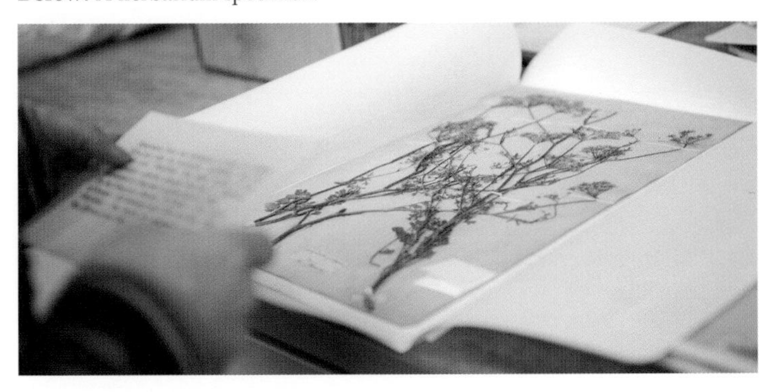

The RBGE Herbarium comprises a collection of more than three million plant specimens with more than 30,000 new items being added each year. In addition to dried plants the Herbarium houses a spirit collection of more than 7,000 specimens. These are primarily fleshy flowers, fruits and fungi that are not suitable for mounting on a herbarium sheet so are preserved in alcohol. There is also a carpological collection made up of seeds and fruit, as well as any plant parts which are too large to be incorporated as a standard herbarium sheet. The RBGE Library is Scotland's national collection of botanical and horticultural literature with more than 70,000 books, large collections of original artworks, manuscripts and other archival materials dating back to the 15th century.

RBGE was founded in 1670 as a physic garden by two friends – Dr Robert Sibbald, the first Professor of Medicine at the University of Edinburgh, and Dr Andrew Balfour, one of the founders of the College of Physicians. In 1676, James Sutherland was appointed by the Town Council to run the Garden and began to teach students at the University of Edinburgh; he was later appointed to the newly created Chair of Botany in 1695. These links between RBGE and the University were to persist until 1956, when the joint posts of Regius Keeper of RBGE and Professor of Botany were finally separated.

Below: Botanical Society of Edinburgh membership ballot box.

The origins of the Herbarium and Library are slightly more recent and can be traced back to a meeting that took place at 15 Dundas Street in Edinburgh on the evening of 8 February 1836. This meeting led to the creation of the Botanical Society of Edinburgh, with one aim being the formation of a public Herbarium and Library.

By the end of the Society's first year there were more than 30,000 specimens of British plants, and a similar number of foreign plants, in the new Herbarium. By 1839, the collection had grown to 150,000 specimens and the Society was faced with the problem of finding somewhere to accommodate it. On 30 July 1838, a Memorial and Petition was presented to the Patrons of the University of Edinburgh which proposed that the Society's collection, already stored in the same rooms as the University Herbarium, should be merged with the University's collection. The proposal was agreed and the amalgamated Herbarium remained at the University until 1863, when, with the holdings continuing to grow, the foreign collections were re-housed at RBGE in what was originally the exhibition hall of the Royal Caledonian Horticultural Society, with the British section following shortly afterwards.

Visitors to the RBGE Library today might be surprised to discover that the Garden managed to get through its first 200 years without an official library. The Regius Keepers bought books when required from their own income and when they left they generally took these collections with them. This all changed on 23 April 1872, when William Craig, the Chairman of the Library Committee of the Botanical Society of Edinburgh, wrote to John Hutton Balfour, the Regius Keeper at the time: "The Society will hand over to the Government the entire Library [about 1,000 volumes], and continue to send any botanical works which they may from time to time receive, on the understanding that the Government will provide for their accommodation and keeping …". The Society's proposal was strongly supported by Balfour and on 30 September that year he received confirmation from HM Office of Works in Edinburgh that the building of an extension to the Herbarium to accommodate the Society's library had been approved.

Above: RBGE Herbarium and Library building during construction.

Unfortunately, the extension that was built was not actually big enough to house the whole of the collection, with the result that, as Balfour noted in his report for 1878, "the books are scattered through various rooms, rendering the consultation of them by the Regius Keeper and Garden visitors very inconvenient". There were similar problems with the Herbarium which, by 1960, had expanded to include more than one and a half million specimens and had outgrown the original space, overflowing into four additional huts and other unsuitable buildings. A new building to re-house both collections was urgently required. The new Herbarium and Library building was duly designed by Robert Saddler (1912–2008) of the Directorate of Works, Ministry of Public Building and Works. Following approval of the design by the Royal Fine Art Commission for Scotland, construction began in June 1962 on a site just south of the Botany building (now known as the Balfour Building) at a cost of £250,000, and the building opened two years later.

While the RBGE Herbarium and Library contain many historical objects, they are very much working collections. They provide a unique research facility that is used both by scientists based at the Garden and by researchers from around the world in exploring, explaining and conserving the world of plants. Knowledge of plants is crucial in understanding many of the challenges facing mankind, including climate change, the loss of biodiversity, food security, the development of new medicines and many more. Behind each new discovery in these fields is a plant that has a name and behind each of those names is a herbarium specimen. Taxonomy, the science of classifying and naming the natural world, is at the heart of the work carried out by RBGE and would be impossible without the Herbarium and Library.

Accumulated over a period of more than 300 years, the RBGE collections provide an important historical perspective, enabling researchers to compare the biodiversity in certain areas over long periods of time, providing invaluable information on change. For example, the collections include species of moss that were growing before, during and after the Industrial Revolution; these can be used to assess levels of atmospheric pollution. The collections are also important as sources of information about the social and economic conditions that RBGE staff experienced. Plant collectors, such as George Forrest and Joseph Rock, were often the first westerners to visit certain parts of the world and their diaries and collecting books provide a unique historical insight into life in these countries.

Without a doubt the hardest part of putting this book together has been deciding what to leave out. Making this selection from the vast collections was an enormous challenge, albeit a hugely enjoyable one. So, what counts as a treasure? The items selected all have a story to tell, whether tales of daring collectors; scientific discoveries and those who made them; plants and their importance for the economy; or cautionary tales of risks to biodiversity and the challenges that we might all face as a result. Many of the objects were just too beautiful to exclude, while others, though possibly less obviously attractive, help to tell the story of the collections and of the Royal Botanic Garden Edinburgh. We hope that you enjoy our final selection.

David Harris – Herbarium Curator
Lorna Mitchell – Head of Library Services

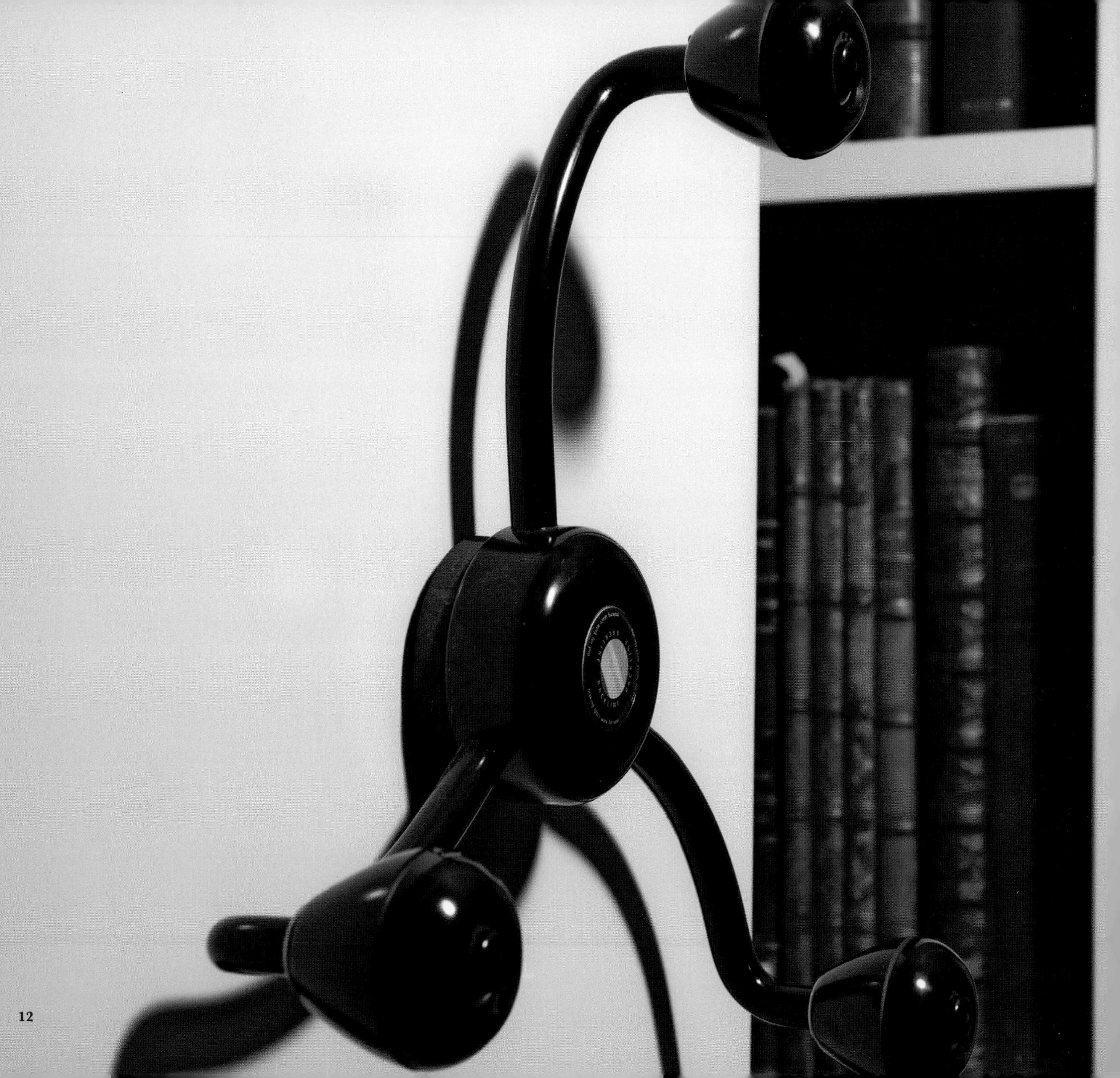

RBGE History: Places and People

For over 300 years, as plants have been discovered around the world, staff at the Royal Botanic Garden Edinburgh have grown them, studied them, taught others about them and inspired generations of botanists and artists to continue to explore, explain and conserve the world of plants. This work continues today as botanists and horticulturists at RBGE work to push science forward. This chapter celebrates the work of those who began our important collections, and those who significantly developed them through time.

Cancer Bush

Sutherlandia frutescens

James Sutherland (c.1638/9–1719) was appointed as the first Regius Keeper of the Physic Garden that became the Royal Botanic Garden Edinburgh, having already become the University of Edinburgh's first Professor of Botany. His skill in the cultivation of medicinal plants had brought him to the attention of Robert Sibbald (1641–1722) and Andrew Balfour (1630–1694), founders of RBGE on sites at the Palace of Holyroodhouse and Trinity College. The genus *Sutherlandia* was named in 1812 by the great Scottish botanist Robert Brown (1773–1858), as it had been grown by Sutherland in the Physic Garden, and listed under the name *Colutea frutescens* in his catalogue of the Garden, the *Hortus Medicus Edinburgensis* of 1683 (p. 15). One of the three collections on this sheet was collected at RBGE in about 1825. **EH**

Hortus Medicus Edinburgensis

Hortus Medicus Edinburgensis, or A Catalogue of the Plants in the Physical Garden at Edinburgh; containing their most proper Latin and English names; with an English alphabetical index, published in 1683, is the earliest botanic garden catalogue printed in Scotland. It appeared only thirteen years after the foundation of the Physic Garden, which was the original incarnation of today's RBGE. In his introduction to the work James Sutherland, the first Royal Botanist in Scotland, discussed his collection plan for the garden: "It having been my Businesse these seven years past, wherein I have had the Honour to serve the City as Intendant over the Garden, to use all Care and Industry by forraign Correspondence to Acquire both Seeds and Plants from the Levant, Italy, Spain, France, Holland, England, east and west Indies; and by many painful Journeys in all the Seasons of the year, to recover whatever this Kingdom produceth of Variety, and to cultivate and preserve all of them with all possible Diligence." The catalogue classified plants as either medicinal (Offic.), annual (Ann.) or found in Scotland (S.), and included cultivated, wild and garden species. Over 2,000 species and varieties were listed. The RBGE Library holds five copies of the book, two of which belonged to previous Regius Keepers. This copy has a fine-tooled leather binding from the early 18th century which uses a small thistle motif in the design. **GH**

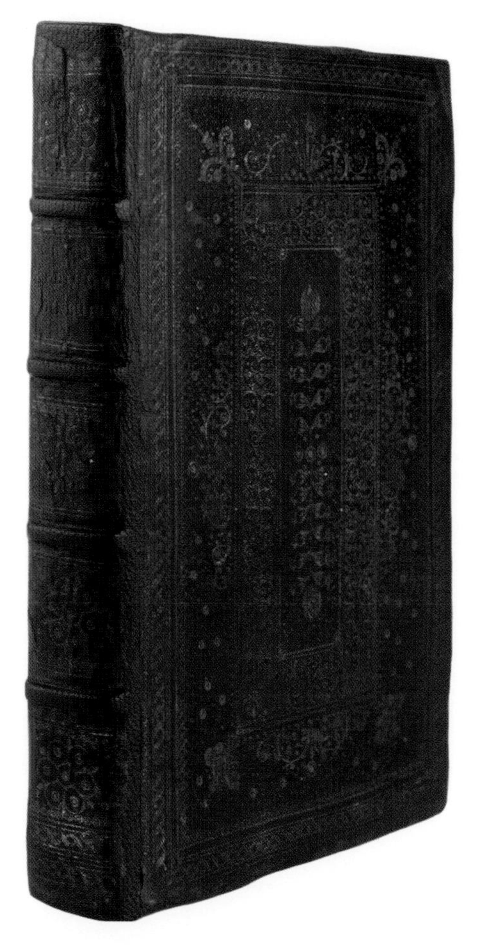

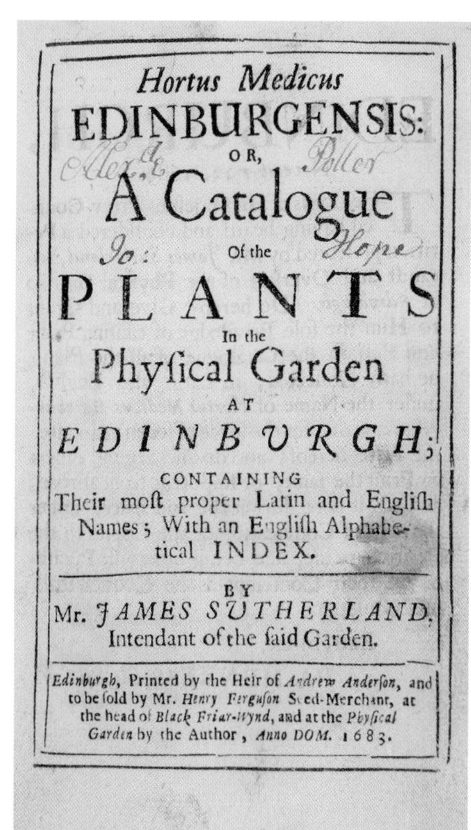

Hortus Medicus
EDINBURGENSIS:
OR,
A Catalogue
Of the
PLANTS
In the
Physical Garden
AT
EDINBURGH;
CONTAINING
Their most proper Latin and English Names; With an English Alphabetical INDEX.
BY
Mr. JAMES SUTHERLAND,
Intendant of the said Garden.

Edinburgh, Printed by the Heir of *Andrew Anderson*, and to be sold by Mr. *Henry Ferguson* Seed-Merchant, at the head of *Black Friar-Wynd*, and at the *Physical Garden* by the Author, *Anno DOM.* 1683.

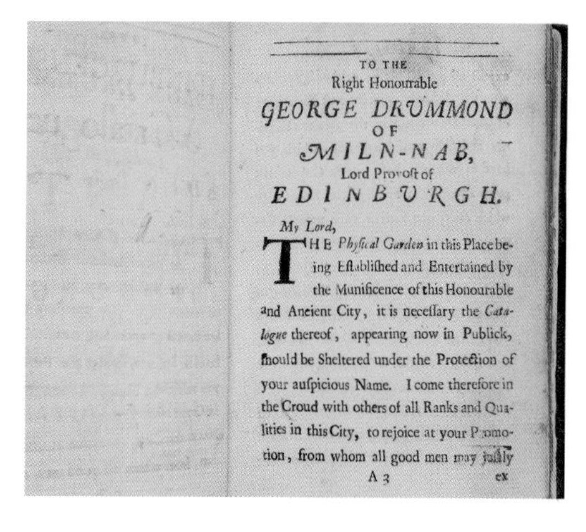

TO THE
Right Honourable
GEORGE DRUMMOND
OF
MILN-NAB,
Lord Provost of
EDINBURGH.

My Lord,

THE *Physical Garden* in this Place being Established and Entertained by the Munificence of this Honourable and Ancient City, it is necessary the *Catalogue* thereof, appearing now in Publick, should be Sheltered under the Protection of your auspicious Name. I come therefore in the Croud with others of all Ranks and Qualities in this City, to rejoice at your Promotion, from whom all good men may justly

A 3 ex

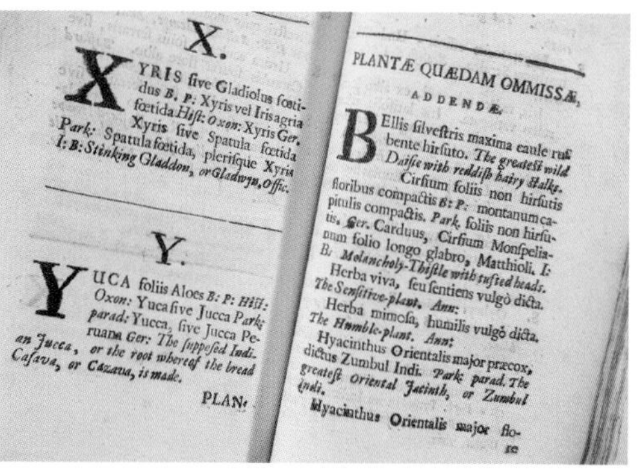

X.

XYRIS sive Gladiolus fœtidus *B. P:* Xyris vel Iris agria fœtida *Hist: Oxon:* Xyris *Ger.* Xyris sive Spatula fœtida *Park:* Spatula fœtida, plerisque Xyris *I: B: Stinking Gladdon, or Gladwyn, Offic.*

Y.

YUCA foliis Aloes *B: P: Hist: Oxon:* Yuca sive Jucca *Park: parad:* Yucca, sive Jucca Peruana *Ger: The supposed Indian Jucca, or the root whereof the bread Casava, or Cazava, is made.*

PLAN-

PLANTÆ QUÆDAM OMMISSÆ, ADDENDÆ.

BEllis silvestris maxima caule rubente hirsuto. *The greatest wild Daisie with reddish hairy stalks.*

Cirsium foliis non hirsutis floribus compactis *B: P:* montanum capitulis compactis. *Park.* foliis non hirsutis, *Ger.* Carduus, Cirsium Monspelianum folio longo glabro, Matthioli. *I: B: Melancholy-Thistle with tufted heads.*

Herba viva, seu sentiens vulgò dicta. *The Sensitive-plant. Ann:*

Herba mimosa, humilis vulgò dicta. *The Humble-plant. Ann:*

Hyacinthus Orientalis major præcox, dictus Zumbul Indi. *Park: parad. The greatest Oriental Jacinth, or Zumbul Indi.*

Hyacinthus Orientalis major flore

Latin *Herbarius*

Printed in Louvain in 1485/6 by Johann Veldener, this herbal is the oldest item in the Library collection. Compiled by an anonymous author it brings together existing texts from medieval, classical and Arabic authors with 150 wood-block illustrations and was intended to provide a guide to medicinal plants for those with no access to a physician. A manuscript inscription at the front of the volume indicates that "Latterly it had belonged to Mr James Sutherland, the first Keeper of the Edinbr. Botanic Gardens; afterwards [1707] to the advocates Library". The book made its way back to the Garden in 1900 when it was purchased at the sale of the library of James Hardy. **LM**

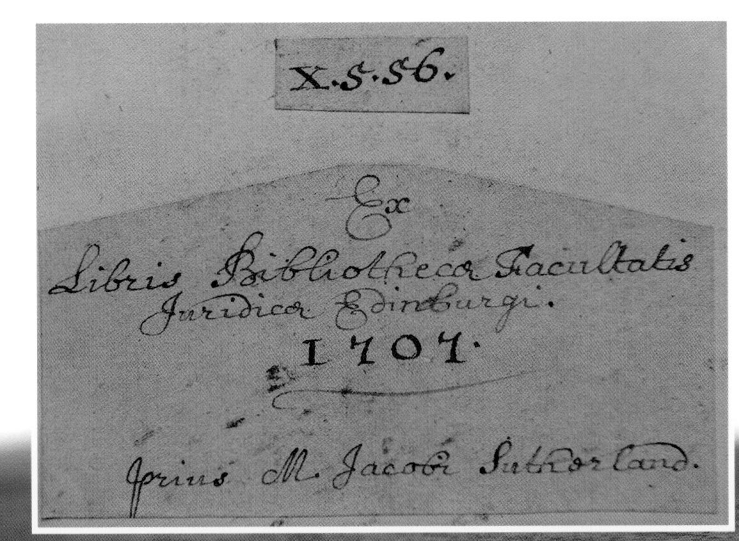
X.5.56.
Ex
Libris Bibliothecæ Facultatis
Juridicæ Edinburgi.
1707.
prius M. Jacobi Sutherland.

Cape Myrtle

Myrsine africana

Collected over 300 years ago, in 1697, at the Cape of Good Hope by Alexander Brown, this is the oldest specimen so far found in the RBGE Herbarium. Brown was a ship's surgeon who travelled to the Far East from 1692 to 1698. The Cape was a regular stopover during these long voyages and Brown was one of many botanists who took the opportunity to collect specimens of its rich flora. This specimen was sent to Charles Dubois (1658–1740), cashier-general of the East India Company and Fellow of the Royal Society, who was an avid collector of shells, fossils, coins and plant specimens. His herbarium, about 13,000 specimens in all, is now largely held at Oxford University Herbaria, but in the past a small number of duplicates were sent to Edinburgh. **HA**

Birdseye View of the Leith Walk Garden

In 1761, John Hope (1725–1786) took up the posts of King's Botanist in Scotland, Regius Keeper of the Botanic Garden and Professor of Botany and Materia Medica at the University of Edinburgh. He inherited the small, overcrowded Physic Gardens in the grounds of Trinity Hospital and at Holyrood, and determined to move them to a new, more spacious (5-acre, approximately 2 hectare) rural site on Leith Walk. Using contacts with the botanically minded Earl of Bute, he obtained a grant from the Treasury in 1763, and plans for the gardener's house and conservatories from his old school friend John Adam.

Although drawn six years before William Crawfurd's plan (p. 20) of Hope's garden, this watercolour shows exactly the same features. It is taken from an imaginary viewpoint hovering somewhere above the south-east side of Leith Walk – in the background are the Firth of Forth and Largo Law. The drawing is one of two commissioned by Hope in 1771 from the young Jacob More (1740–1793), for which the artist was paid one guinea for the pair. At this point More, having trained with the Norie family and Alexander Runciman, was painting scenery for the Theatre Royal, but he would shortly leave Scotland for Rome, where he made his name with oil paintings of idealised classical landscapes inspired by Claude Lorrain. Here he shows Hope's recent creation, a unique combination of miniature landscape garden and traditional botanic garden, which fulfilled both teaching and research functions. The drawing shows the original level of Leith Walk, which was later raised to the height of the first floor of the gardener's cottage. **HN**

Plan of the Leith Walk Garden

This plan shows RBGE on its Leith Walk site at the height of John Hope's tenure of the joint posts of Professor of Botany and Regius Keeper of the Garden. It was drawn by the surveyor William Crawfurd and is dated 6 September 1777, when the Garden was about 12 years old. It shows several of the unique features of Hope's garden, such as the central placing of the 140-foot-long (63 metre) range of greenhouses, and the informal layout of the main section of the Garden with curving walks and similarities to contemporary landscape gardens. The formal area, the 'Schola Botanica', more typical of traditional botanic garden design, is restricted to the northern part, outside which is the field where Hope grew his precious rhubarb (p. 21). Around the pond is a narrow area of rock-work for the growing of alpine plants. The Garden left Leith Walk in 1822, leaving behind only the gardener's cottage, which is now being rebuilt at the present Inverleith site. **HN**

John Hope's Rhubarb

Rheum palmatum

Eighteenth-century medicine was obsessed with the letting of blood and the purging of bowels. For the latter purpose the medicinal rhubarb was greatly in demand. In 1763, through the Scottish surgeon James Mounsey, who had worked for many years in Russia, John Hope obtained seeds of *Rheum palmatum* (which is native to western China) from the Saint Petersburg Botanical Garden. Hope cultivated 3,000 rhubarb plants in a plot outside the Leith Walk Garden (see p. 20), and distributed it widely to Scottish garden owners, so that requirements for the medicinally active dried root could be met locally without the need for costly imports. This specimen comes from an anonymous 18th-century collection that was incorporated into the RBGE Herbarium in the mid-19th century; although the specimens have no collection details it is possible that this one was collected from one of Hope's Leith Walk Garden plants. The plant is now commonly grown in gardens for its statuesque habit and attractive foliage. **HN**

this is shewen at
the lecture on the
motion of the Sap

Experiment on the Motion of Sap

This beautiful drawing, of c.1775, by one of John Hope's gardener-assistants, is of interest from several perspectives. It is actually a highly romanticised adaptation of a rather severe etching in one of Hope's best-loved reference books, Stephen Hales's *Vegetable Staticks* (1727). The book describes Hales's experiments on the movement of water and sap in plants, several of which, including the one shown here, were repeated by Hope in his Leith Walk Garden. The outer branches of three willow trees were grafted together and after some years the central tree was excavated – it survived, due to lateral movement of water from the still-rooted flanking trees. Andrew Fyfe (1754–1824) studied drawing at the Trustees Academy under Alexander Runciman, later becoming better known for his anatomical drawings and as a lecturer in anatomy at the University of Edinburgh. Fyfe married Agnes Williamson, daughter of Hope's head gardener, and she also made teaching drawings like this one for use by Hope in his lectures. **HN**

this is shewn at
the lecture on the
motion of the sap

Notes from John Hope's Lectures

None of John Hope's own notes for his influential botanical lectures given between 1761 and 1786 survive, and only five sets taken down by students who attended still exist. The most intriguing of these sets was made by Francis Buchanan-Hamilton (1762–1829), who attended the lectures at the Leith Walk Garden in 1781. Buchanan-Hamilton became an East India Company surgeon, and as a surveyor and compiler of 'statistical' information he made major contributions to numerous fields of study in the Subcontinent – including botany, zoology, agriculture and archaeology. He was the first to make botanical collections in Nepal, of which he made a manuscript Flora that was never published. RBGE is currently involved in writing what will therefore be the first comprehensive account of the plant life of Nepal. Buchanan lent his volume of lecture notes to a shipmate in 1785 after which it fell into the hands of Tipu, Sultan of Mysore, who had it bound for his library. At the fall of Tipu's capital of Seringapatam in 1799 it was retrieved by a soldier and returned to Buchanan-Hamilton, ending up in the RBGE Library only in 1966. **HN**

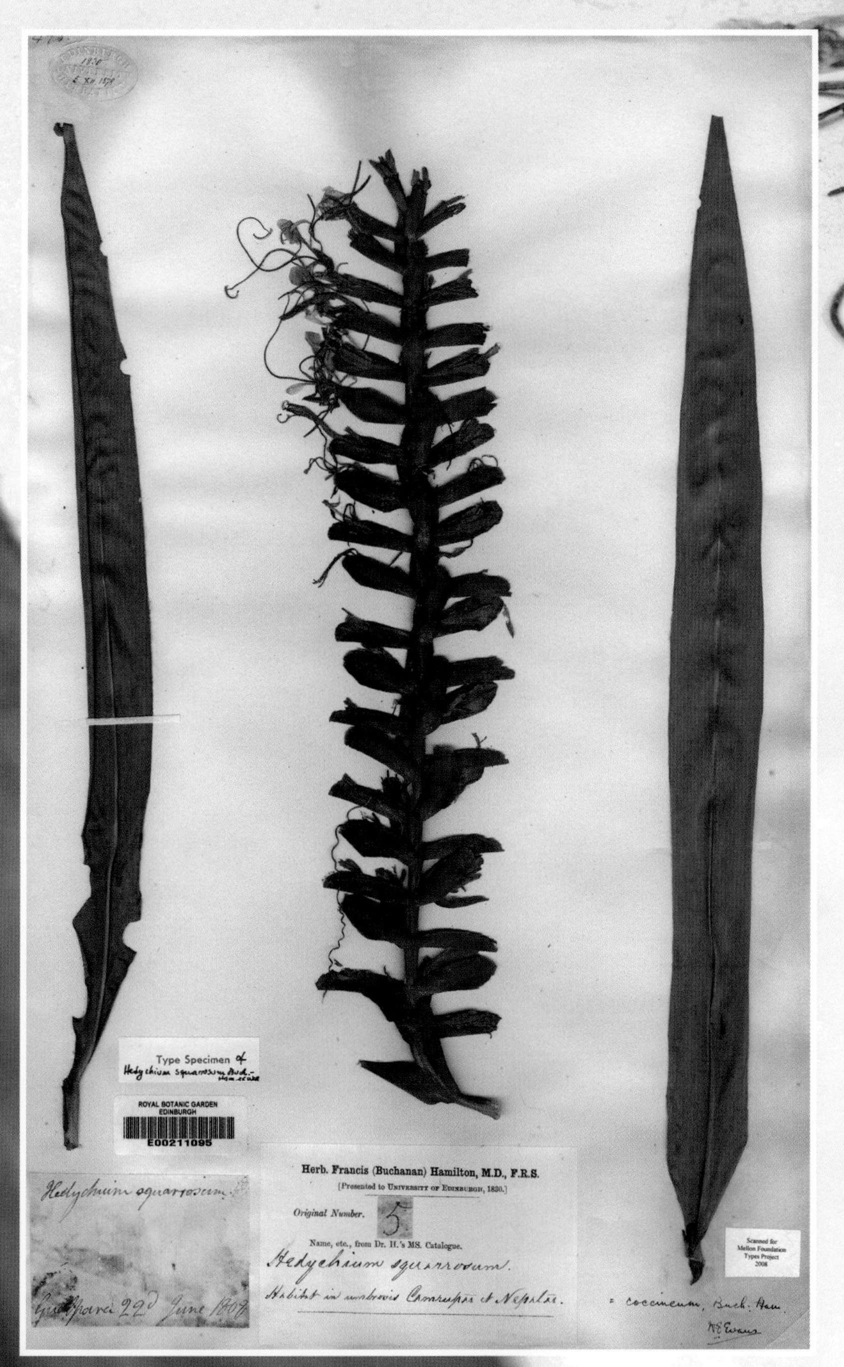

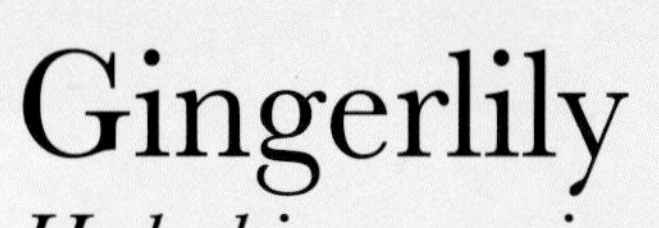

Gingerlily

Hedychium coccineum

This specimen is a tall, stately, yellow-flowered member of the ginger family (Zingiberaceae) gathered by Francis Buchanan-Hamilton from shaded forests near the Indian village of Gualpara on the southern border with Nepal. At the time he thought it was new and named it '*Hedychium squarrosum* B.' on his field ticket, but it turned out to be the same as a plant he had collected six years earlier in Nepal, and described by James Edward Smith in 1811 using the name on Buchanan-Hamilton's Nepalese specimen. Buchanan-Hamilton was one of the most accomplished of the East India Company's naturalist-surgeons, and a close friend of William Roxburgh (1751–1815), Superintendent of Calcutta Botanic Garden. Both these Scotsmen had been taught botany by John Hope at RBGE as part of their medical degree at the University of Edinburgh. Buchanan-Hamilton sent many seeds and living plants to Calcutta and his name features prominently in Roxburgh's catalogue of the garden, *Hortus Bengalensis*. **MW**

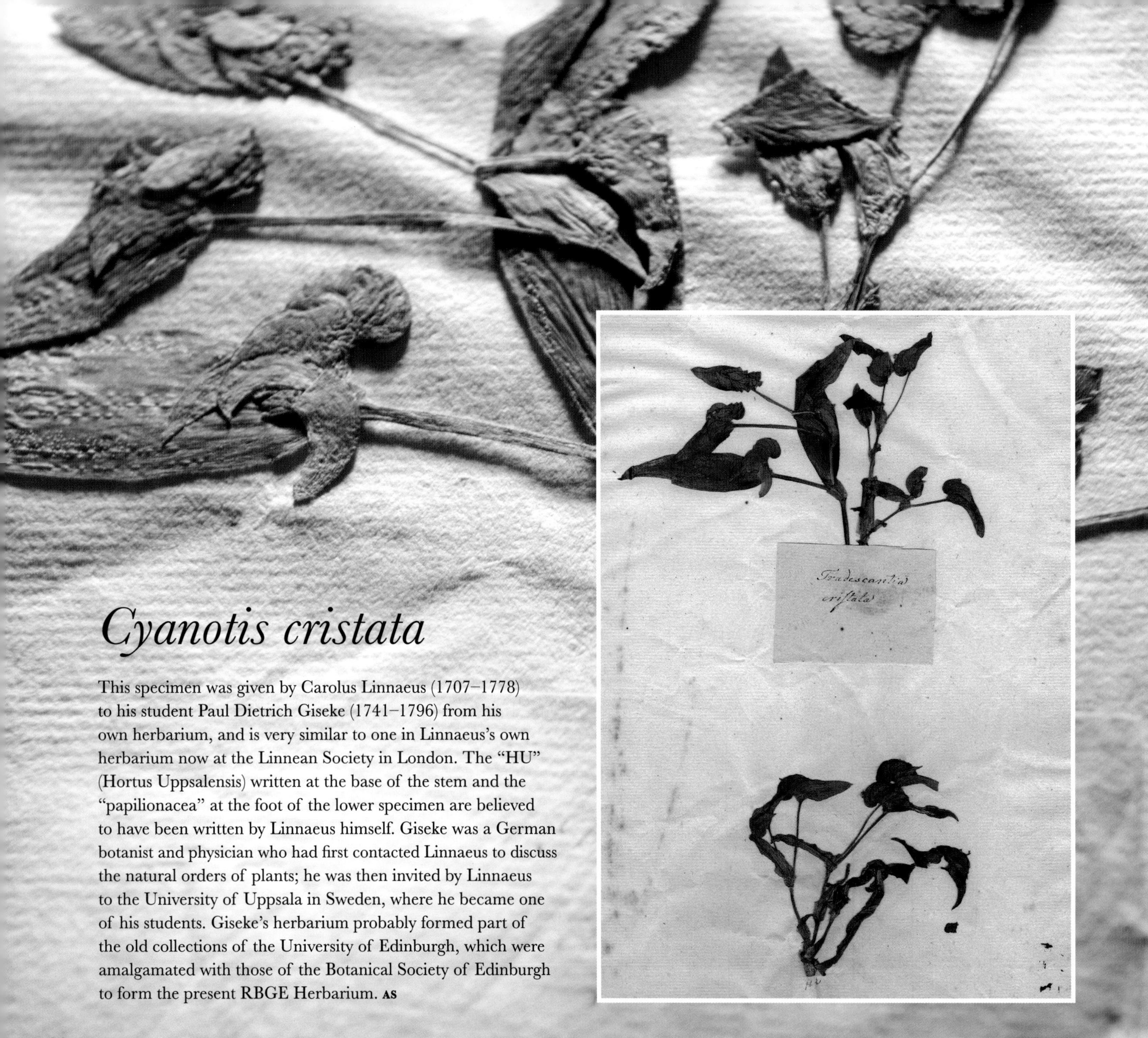

Cyanotis cristata

This specimen was given by Carolus Linnaeus (1707–1778) to his student Paul Dietrich Giseke (1741–1796) from his own herbarium, and is very similar to one in Linnaeus's own herbarium now at the Linnean Society in London. The "HU" (Hortus Uppsalensis) written at the base of the stem and the "papilionacea" at the foot of the lower specimen are believed to have been written by Linnaeus himself. Giseke was a German botanist and physician who had first contacted Linnaeus to discuss the natural orders of plants; he was then invited by Linnaeus to the University of Uppsala in Sweden, where he became one of his students. Giseke's herbarium probably formed part of the old collections of the University of Edinburgh, which were amalgamated with those of the Botanical Society of Edinburgh to form the present RBGE Herbarium. **AS**

Veronicastrum sibiricum

This specimen of the eastern Asian speedwell, *Veronica sibirica*, was collected from the garden of Carolus Linnaeus's summer cottage at Hammarby, near Uppsala, in Sweden in 1771. The species was described in 1762 by Linnaeus in the second edition of his monumental work, *Species Plantarum*. The groundbreaking first edition of this work in 1753 played a major role in the consistent binomial naming of species, which reduced the long, descriptive phrases previously used to just two words – a noun for the genus, and an epithet for the species. Linnaeus gave this specimen to Paul Dietrich Giseke and from him it came to Edinburgh. The Latin writing on the label may be translated as "Given by Linnaeus from his garden at Hammarby". **EH**

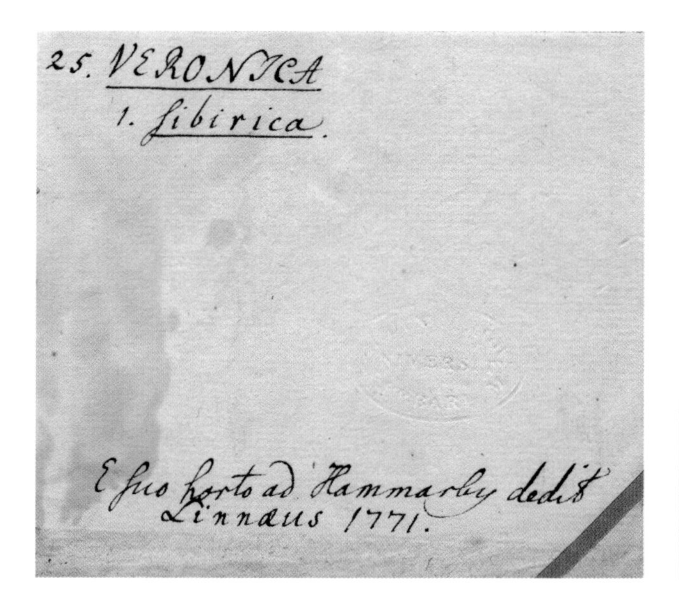

John Hutton Balfour's Teaching Models and Diagrams

Born in Edinburgh, John Hutton Balfour (1808–1884) was one of the Garden's most influential and longest-serving Regius Keepers, holding the post from 1845 until 1879. His time at the Garden was marked by many significant developments but he is particularly remembered as a teacher and as the founder of the Botanical Society of Edinburgh.

A keen advocate of using art to present and explain science, during his time at the Garden he accumulated a collection of 3,681 teaching diagrams that he used in his classes. Sadly a large part of this collection was destroyed in the 1950s and only 368 remain. The majority of the diagrams were the work of Robert Kaye Greville (1794–1866) and John Sadler (1837–1882), Balfour's assistant and later, from 1879 to 1882, the Curator of the Garden.

The botanical models shown here were originally part of a museum of economic botany that was established by Balfour in 1851. Created from papier-mâché, wood, plaster and gelatin, they were produced in Berlin by the Brendel Company, founded by Robert Brendel in Breslau in 1866. Sadly, most of the museum collections were lost in the 1960s during renovations. **LM**

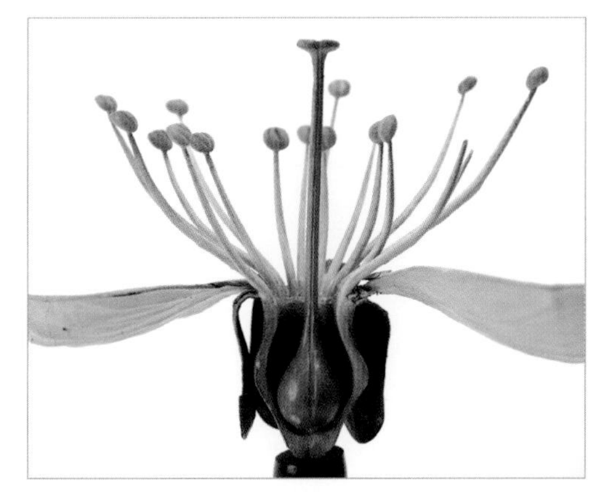

Coco de Mer

Lodoicea maldivica

This object, measuring approximately 35 centimetres across at its widest point, is a single seed from the Coco de Mer palm, the species with the largest seeds in the plant kingdom. Although the palm originates in the Seychelles, its seeds can travel great distances, floating on the sea, to be washed ashore on distant beaches. Early sailors had never seen the plants growing and believed these extraordinary seeds floated up from mysterious forests growing on the sea bed. Their remarkable shape gave rise to legends of fertility. The species is now at risk, being restricted to only two natural populations in the wild. This specimen is one of the few items remaining from the Museum of the Royal Botanic Garden Edinburgh started by John Hutton Balfour. It has been carved and polished to emphasise its voluptuous form. **AS**

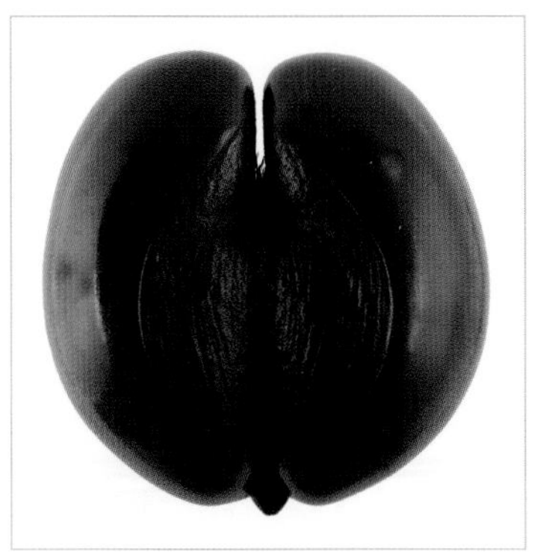

Ordeal Bean

Physostigma venenosum

In 1846, the Scottish United Secession Church set up a mission station in Old Calabar, Nigeria. Several of the missionaries had interests in natural history including the Rev. William Cooper Thomson (nephew of the architect Alexander 'Greek' Thomson) and the Rev. Zerub Baillie, a former student of John Hutton Balfour. One of the missionaries' concerns, which combined moral reform and botany, was a leguminous seed called *eséré*, given as an ordeal to men accused of crimes – if they vomited they were innocent, if they died they were guilty. Specimens were sent to Edinburgh for chemical analysis by Robert Christison (1797–1882), who tested its effects on himself, and for botanical classification by Balfour. With advice from Thomas Anderson and George Bentham, Balfour described the plant as a new species in a new genus (the name referring to the swollen stigmatic appendage), based on Baillie's specimen and a description by Thomson. **HN**

Plant Scenery of the World

In preparation, in Imperial 4to,

A NEW BOTANICAL WORK BY PROFESSOR BALFOUR AND DR. GREVILLE,

WITH CHROMO-LITHOGRAPHIC ILLUSTRATIONS,

THE

PLANT SCENERY OF THE WORLD

A POPULAR INTRODUCTION TO

BOTANICAL GEOGRAPHY.

BY

JOHN HUTTON BALFOUR, A.M., M.D., F.R.SS.L. & E.

REGIUS KEEPER OF THE ROYAL BOTANIC GARDEN, AND PROFESSOR OF BOTANY IN THE UNIVERSITY OF EDINBURGH,

AND

ROBERT KAYE GREVILLE, LL.D., F.R.S.E., M.R.I.A.

MEMBER OF THE IMP. ACAD. NAT. CUR.; CORRESPONDING MEMBER OF THE NAT. HIST. SOC. OF PARIS, AND OF LEIPSIC; OF THE ACAD. OF NAT. SCIENCES, PHILADELPHIA; OF THE LYCEUM OF NAT. HIST., NEW YORK, ETC. ETC.

In an obituary notice for his friend Robert Kaye Greville in the 1866 *Transactions and Proceedings of the Botanical Society of Edinburgh*, John Hutton Balfour gave the history of this subscription work that never reached publication: "A few years ago Dr Greville and I commenced a work, which was to be entitled 'Plant Scenery of the World,' in which the characteristic floras of different regions were to be represented, and a description was to be given of the plants in the landscape. He completed forty or fifty of the coloured plates, and I prepared some of the letterpress, but owing to the difficulty of getting good coloured lithographs – such at least as were reckoned satisfactory by Dr Greville – it was abandoned by the publishers … I still think that the work might be carried on, and that it might prove valuable as illustrating the geographical distribution of plants."

The plan was to issue, on a monthly basis, a total of 30 parts at a cost of 2s each. This was a time when terms such as ecology and phytogeography were coined to describe new interpretations of the natural world. Balfour, as is shown by this prospective project, was a believer in popularising these new developments in understanding the place of plants in particular landscapes. In some ways *Plant Scenery of the World* would have filled the same role for its contemporary audience as natural history television documentaries do today. **GH**

The McNab Family Scrapbooks

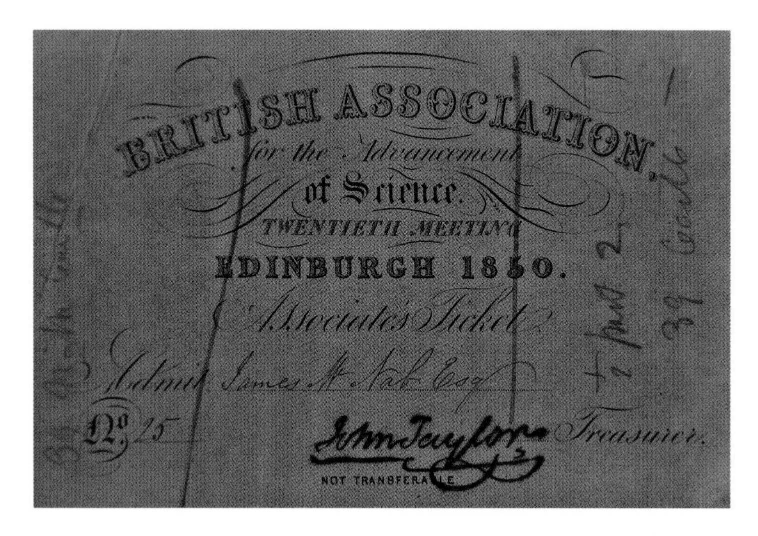
BRITISH ASSOCIATION,
for the Advancement
of Science.
TWENTIETH MEETING
EDINBURGH 1850.
Associate's Ticket
Admit James McNab Esq
No. 25
John Taylor Treasurer.
NOT TRANSFERABLE

Two scrapbooks in the RBGE Archive shed light on the lives of William (1780–1848) and James (1810–1878) McNab, father and son, who between them oversaw the creation and development of the present-day garden at Inverleith, between 1820 and 1878. The Palm Houses and the Caledonian Hall are two tangible parts of their legacy. When James's daughter Susan donated the scrapbooks in 1924, she wrote: "I have no doubt there is a good deal in them of no use but you will be the best judge of that." The passage of almost a century has proved that what she deemed of 'no use' has in fact proved to be of great interest to historians of science, horticulture and social history, and each generation finds new uses for the content. The scrapbook shown here contains advertisements; samples of James's botanical art, including a species of *Polygala* – his first attempt at making an engraved plate; passports; plant lists; testimonial subscription lists; prize certificates; receipts; death notices; and letters from leading botanists and horticulturists, including Robert Fortune, John Claudius Loudon and Joseph Paxton. The second scrapbook contains newspaper cuttings relating to RBGE and proof copies of James's numerous contributions to the horticultural press. **GH**

…r Sir
I have looked over
2 large volumes of my
…her's Botanical album &
…ken out all private papers.
Can you kindly send for them
at your convenience as they are
heavy: They are ready to lift.
I have no doubt there is a
good deal in them of no use but you will
be the best judge of that.
I have still some of the journa…
to look over which you will get later.
With kind regards.
yours sincerely
Susan McNab

two Genttlemans Bill

18 Meals 25c — $4.50
4 Half pint Brandy — 1.00
3 Half pint Wine — 75
5 night Lodging — 1.25
$7.50

Recd. payment in full John Stotler
Alleghaney Sept. 14th. 1834

The Botanical Society Club Archive

The Botanical Society Club was an 'inner circle' of the Botanical Society of Edinburgh. It was formed in 1838 when John Hutton Balfour invited a selection of the original members to his home for dinner on the second anniversary of the Society's foundation. The Club consisted of the 21 original members and met annually on or around the Society's anniversary for supper at one member's house. The Club existed for almost 100 years, each member eventually being replaced by a candidate from the Society on their resignation or death. The menus and discourse of the Club are recorded in two volumes held in RBGE's Archive and seemingly became more elaborate as time went on. Some of the menus are illustrated here, including one produced when archaeological discoveries amongst the Pyramids made Egyptology the height of fashion. **LP**

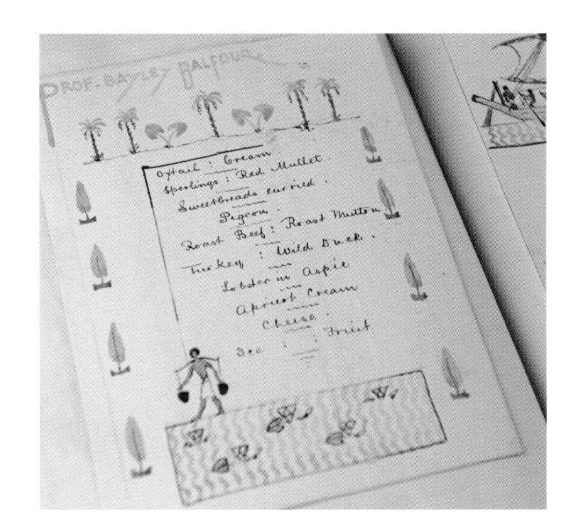

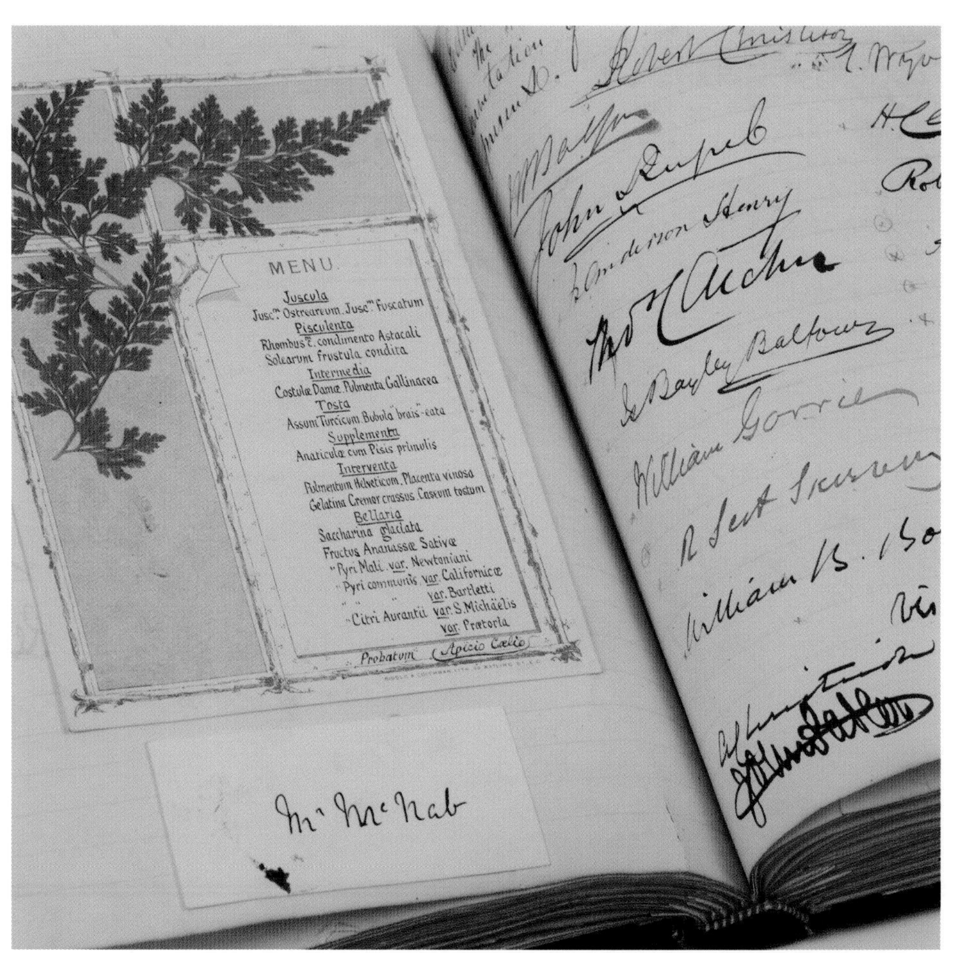

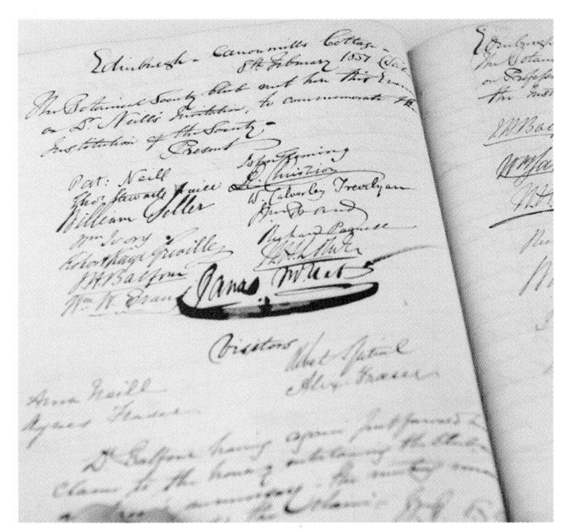

...nounced a fo...
...bers not present. ... Balfour ...
... Dr. John Hutton Balfour ... unwritten law...
... break through their unwritten law...
...toasts at the meeting of the Club – a...
...proposed the health of Mr. Sadler...
...his long connection with the Botanic...
Mr. Sadler made a suitable reply.

Botanical Society Club

1881

Summer Meeting 22nd July

Menu

Soups.

Albert and Hotch Potch

Fish.

Salmon and Fried Haddocks

Entrées.

Lobster Mayonnaise

Pigeons in Truffles

Fillets of Duckling with Green Peas.

Joints.

Roast Lamb

Fowls in Béchamel Sauce and Tongue

Relevés.

Cabinet Pudding – Gooseberry Cream Soufflé

Calf's Foot Jelly – Strawberry Cream

Sallad and Cheese Soufflé

Ices.

Strawberry Cream and Lemon Water

Fruits.

Salix Sadleri

Palm House Architectural Plans

A range of architectural plans are held in the RBGE Archive covering the Glasshouses, the Balfour Building on Inverleith Row and the Library and Herbarium, but the earliest ones in our collection, those of the Palm House built in 1858 and designed by Edinburgh Office of Works architect Robert Matheson, stand out for their aesthetic qualities. Beautifully and meticulously drawn, and coloured with watercolour, they are works of art in their own right. **LP**

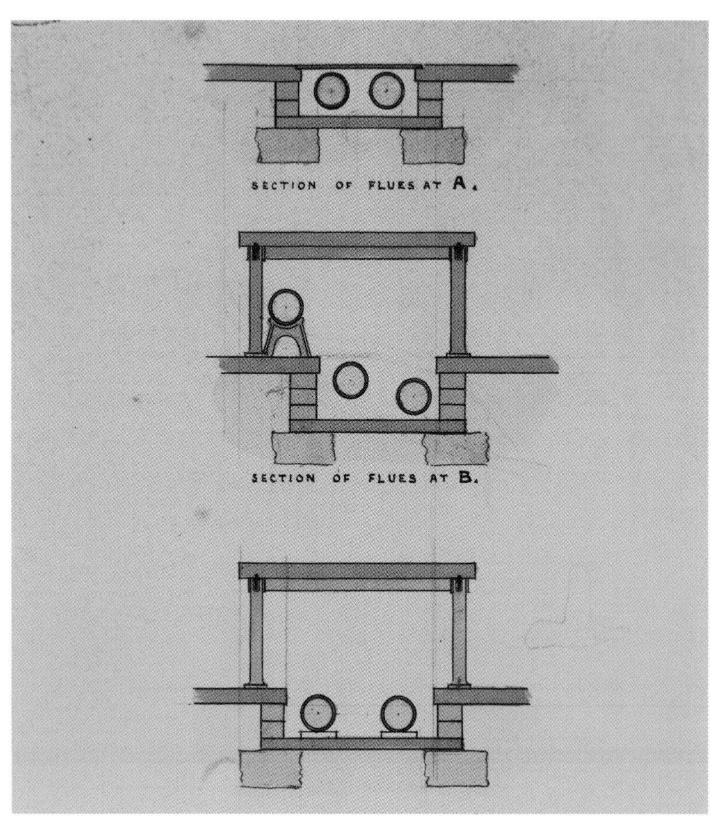

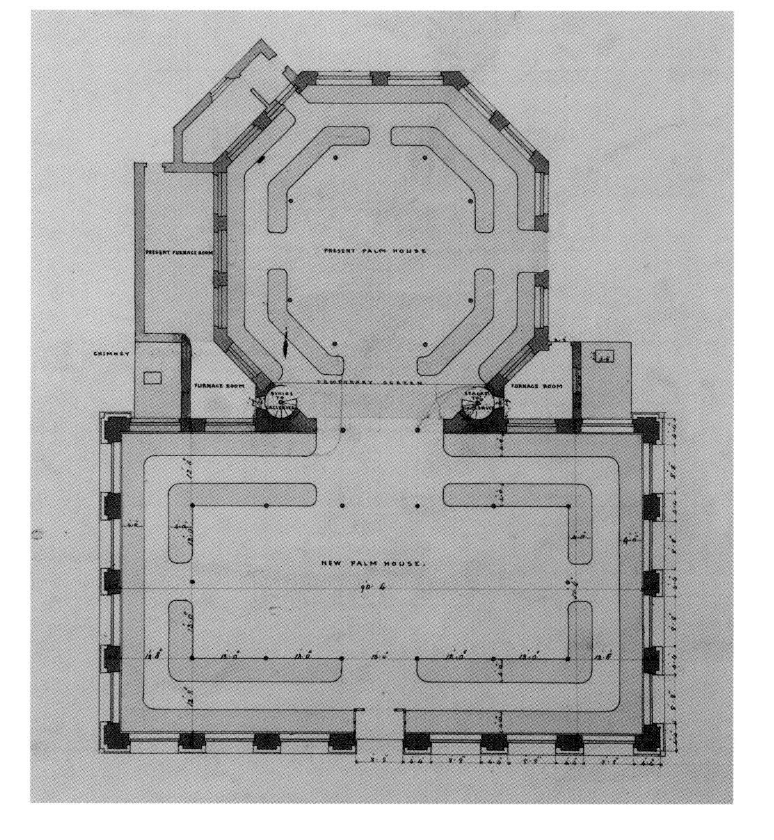

The Original Palm House Photo

Taken in 1855 by Dr James Duncan and printed on paper coated with silver salts, this is the earliest photograph in RBGE's collection. It is of the original Palm House built on the Inverleith site in around 1834, with the west end of the Glasshouse Range visible in the foreground. Palms can be clearly seen protruding through the Palm House roof, enhanced by strokes of ink which are the same shade as the original print would have been. This image was successfully used to lobby Parliament to obtain funding for a larger, taller Palm House. The result was the current Temperate Palm House, which was opened in 1858 and is situated in front of its earlier counterpart. The note which accompanies the print was written by the Temperate Palm House's architect, Robert Matheson, and announces the success of the funding bid. At the top it is observed that the note arrived at the Garden just as the Edinburgh Castle guns were firing to mark the end of the Crimean War in 1856. **LP**

31. March 1856. This note reached the [illegible] at ¼ past 12 o'clock while the Castle guns were firing for the Peace—

Monday

My dear Sir,

I am glad to inform you that the Estimates for the New Palm House have been approved of — namely Messrs Beattie & Son, Shotts Iron Foundry & Russell.

Yours truly

R Matheson

P.S. I met Dr Balfour at the station this morning & informed him

Bleeding Heart Vine

Clerodendrum thomsoniae

The identity of the woman after whom this species was named was a mystery for many years. It was known that a young woman had married a divinity student named Thomson and that she had tragically died in West Africa. During her fever she had constantly asked after a beautiful pink-flowered plant that she had discovered and planted in her garden. Her grieving husband sent the plant to John Hutton Balfour at RBGE, who named it in her honour. This was all that was known of the story until 1946 when the daughter of Mr Thomson and his second wife heard of the mystery. She revealed that his first wife was Mary Stewart, the daughter of a farmer in the Scottish Borders. Her name was given to a plant which has since become known to gardeners around the world. This specimen was made from the original plant which Mary Thomson planted in her garden and which her husband then sent to Balfour. **EH**

Royal Caledonian Horticultural Society Certificate

The Royal Caledonian Horticultural Society (RCHS) was formed in 1809 as a medium for gardeners to discuss and experiment with horticultural techniques and to display their successes in regular shows. A certificate was designed in 1851 which was awarded to winners of the prizes and medals the Society established. This is the original copy, beautifully decorated with floral motifs. It also shows the two founders, Dr Andrew Duncan (1744–1828) and Patrick Neill (1776–1851), and the Caledonian Hall, where shows and meetings were held, now adjacent to the RBGE Rock Garden, but originally at the centre of the RCHS's Experimental Garden. This image is the only one we have that shows the Society's glazed Winter Garden, situated to the south of the Caledonian Hall and long since demolished. **LP**

Garden Map

When RBGE moved its Edinburgh Garden from its site on Leith Walk to its current position in Inverleith in the early 1820s, the area it occupied was only a fraction of the size it is now, extending to just west of the Palm Houses and south of the Pond. It began to expand in the 1860s, a significant addition being the ground of the Royal Caledonian Horticultural Society's Experimental Garden to the south. This area was developed into the Rock Garden by the Curator at the time, James McNab.

This incredible map, produced c.1870 and measuring a not inconsiderable 206 x 130.5 centimetres, shows the Garden shortly after the new land to the south had been acquired. Vestiges of the wall which had once separated the two gardens can be seen, as well as the layout of plants and buildings at the time; the arrangement of these reflected the Garden's prime purpose as a teaching facility. **LP**

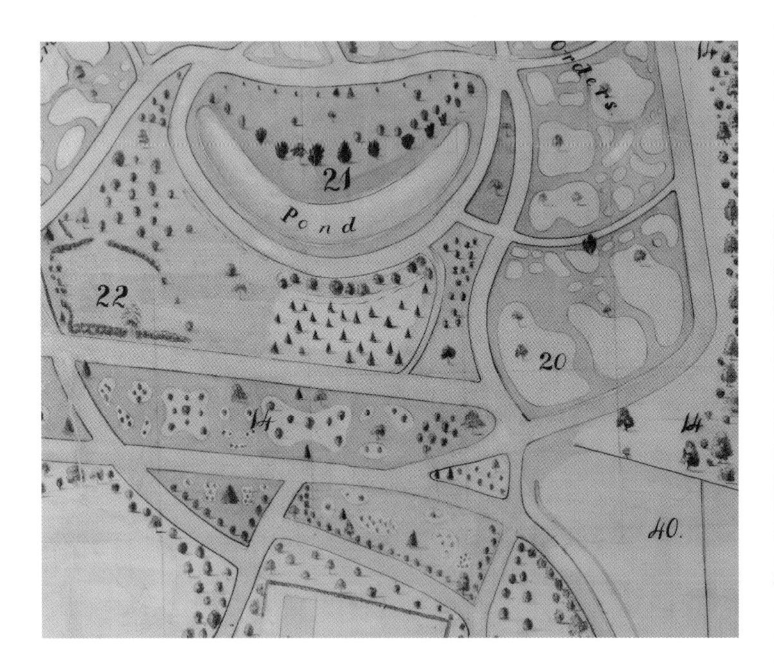

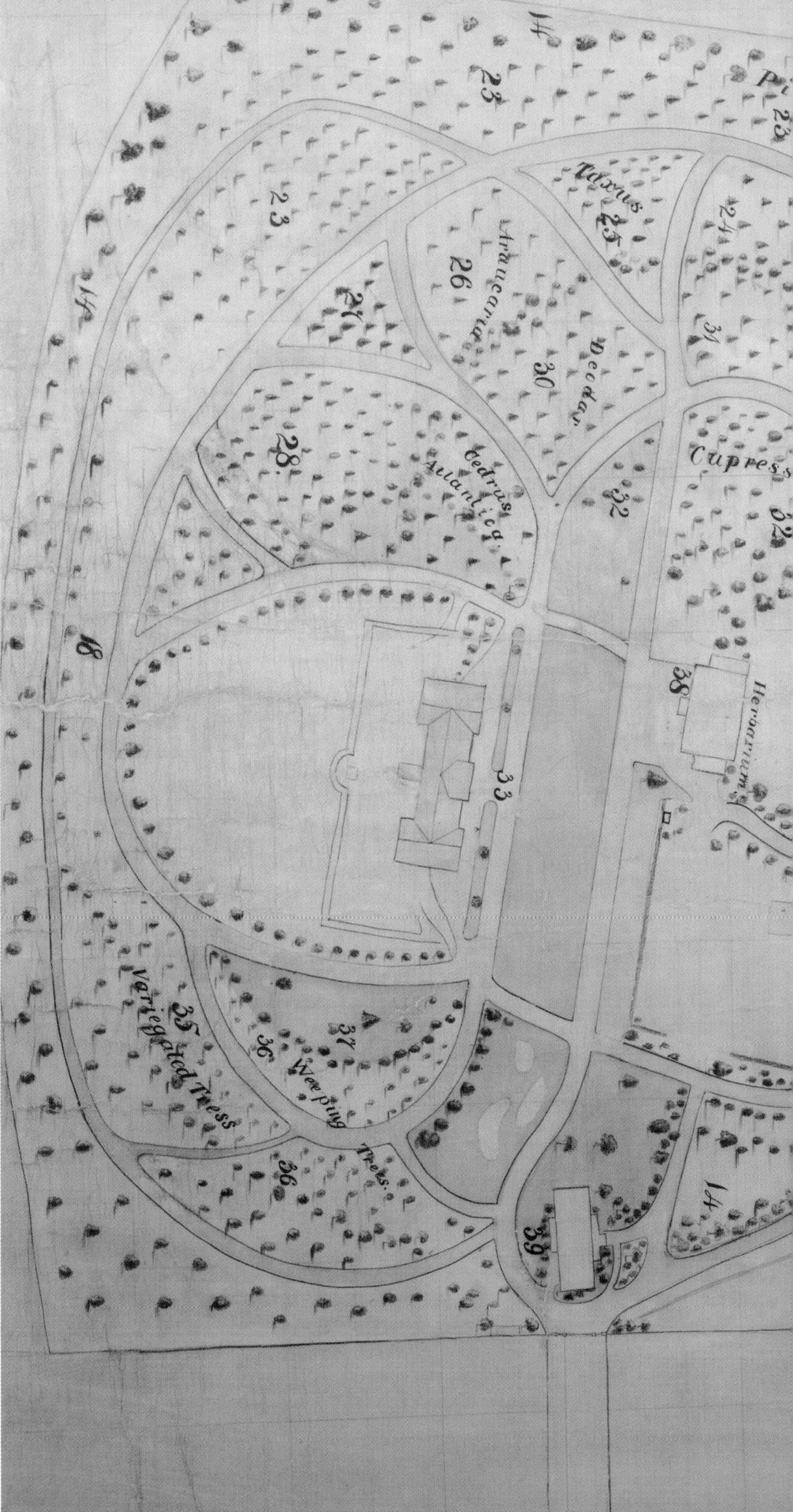

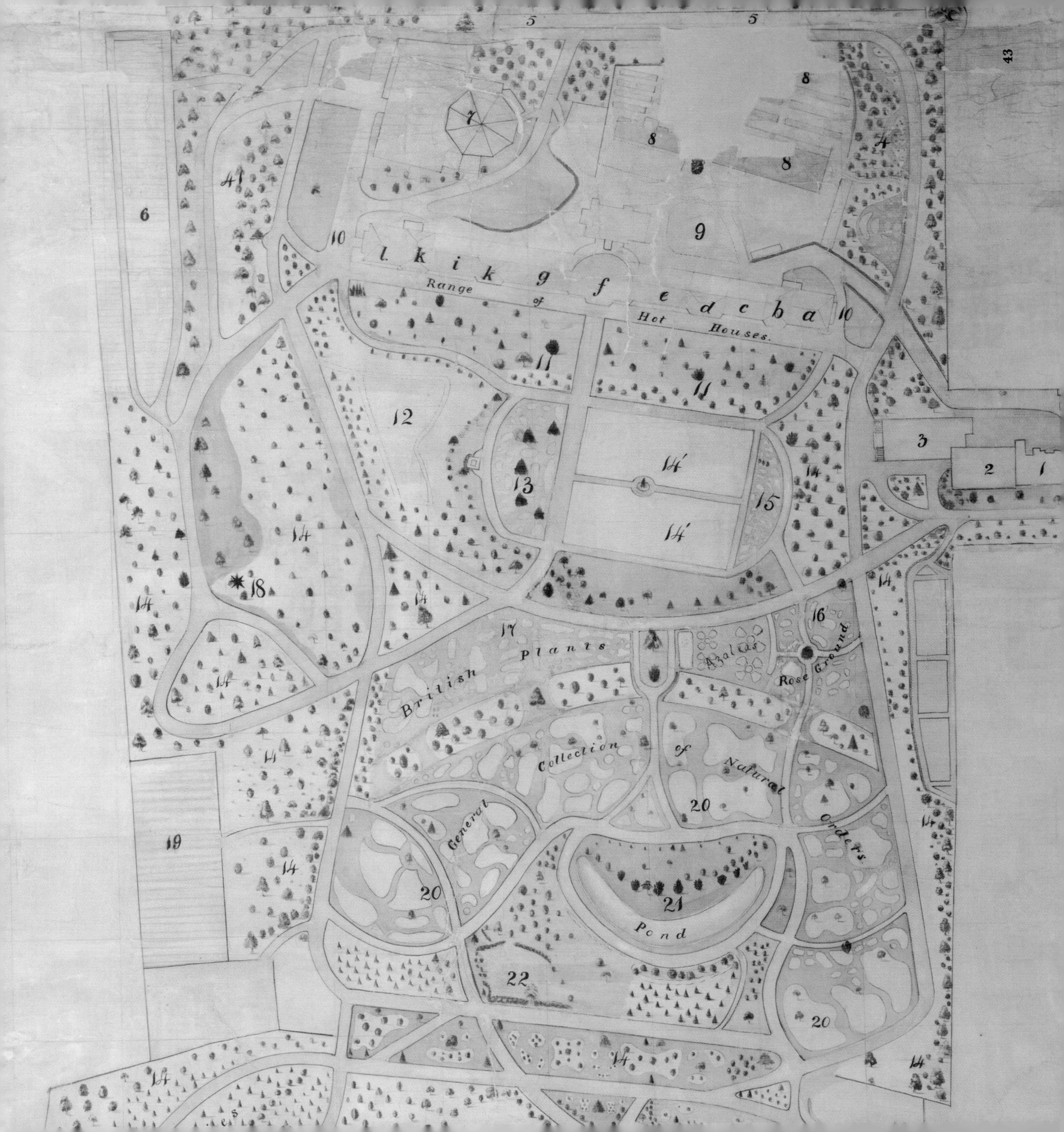
l k i k g f e d c b a
Range of Hot Houses
British Plants
General Collection of Natural Orders
Azaleas
Rose Ground
Pond
1
2
3
4
5
6
7
8
9
10
11
12
13
14
15
16
17
18
19
20
21
22

Morphological Drawings by Alexander Dickson

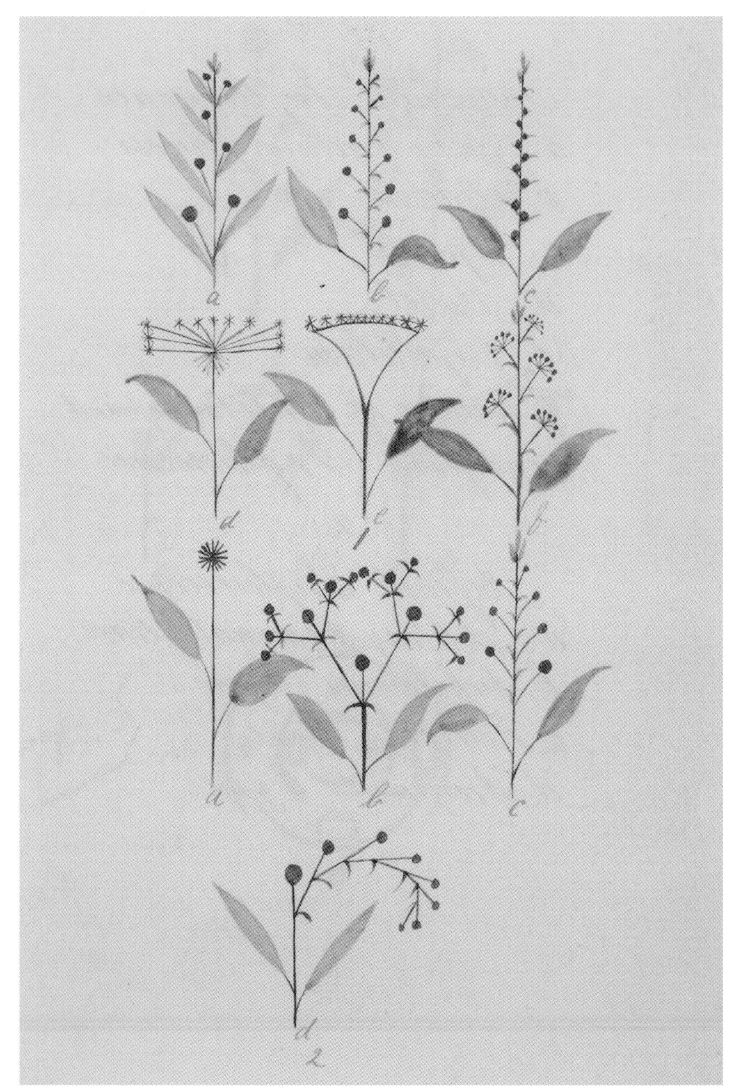

Alexander Dickson (1836–1887) was the Garden's Regius Keeper from 1880 until his sudden death in 1887. Very much a research scientist, and a trained doctor like his predecessors, he was much loved by his students, despite enjoying teaching less than other aspects of his job. Dickson preferred not to employ the large hanging teaching diagrams used by his predecessor John Hutton Balfour; instead he produced drawings himself on the Lecture Theatre's blackboard each day – he felt that students were more likely to pay attention to, and copy down, something that would be erased at the end of the day. It is easy to see from this series of drawings from the Archive what a skilled draughtsman Dickson was. **LP**

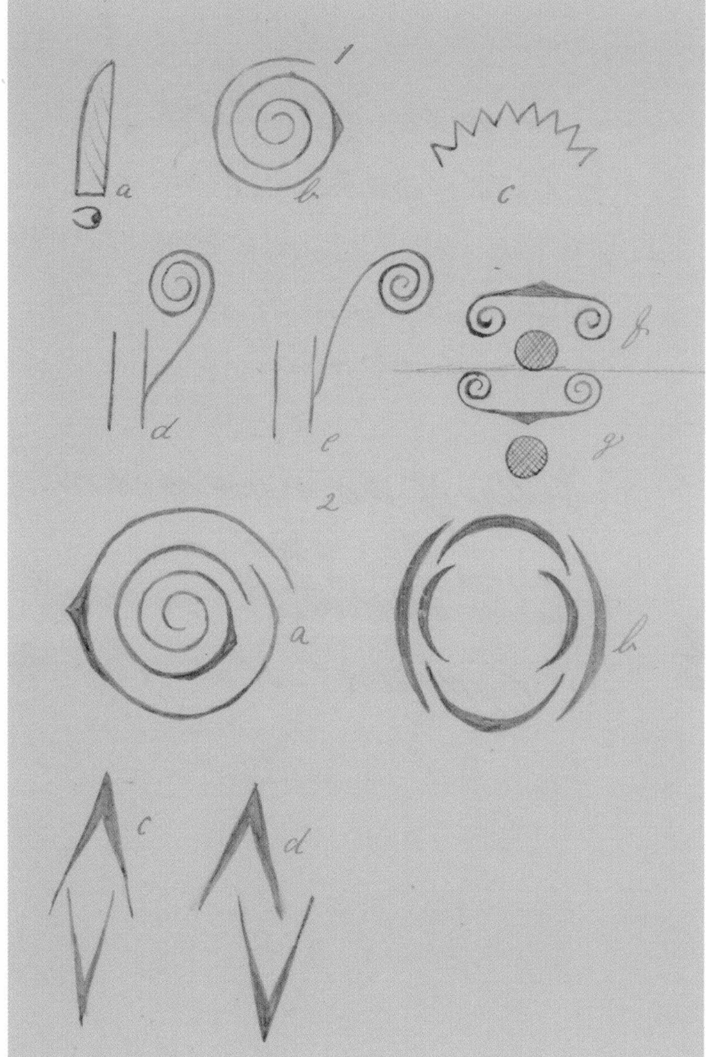

Robert Moyes Adam's Photographs

Robert Moyes Adam (1885–1967) was RBGE's first official photographer, being appointed 'Assistant in the Studio' by Regius Keeper Isaac Bayley Balfour (1853–1922) in 1915. He held the post until his retirement in 1949. Prior to this appointment much of RBGE's early photographic record had been compiled by senior members of the Horticulture Department; indeed Adam himself began his career at RBGE as an Assistant Head Gardener in 1903. During his career Adam photographed many scenes around the Garden, including the glasshouse ranges. He accumulated an impressive collection of plant portfolios, images both of plants growing in the Garden and of potted plants and cuttings in his studio. In particular he photographed many of the Asian *Primula* and *Rhododendron* species that the RBGE taxonomists were concentrating on at that time. **LP**

THE FOREST HERBARIUM (BKF)
ROYAL FOREST DEPARTMENT
FLORA OF THAILAND
Botanical Name
Collector
No.
ABDUL WAHAB B. ABDUL SA
PUNIRAN B. MARJAN
ABDUL B. SAMAD
SUKOR B. KARTA
PETRUS KAPANG
HASHIM B. WAHAD
Botanical Na
Collector
No.

Exploration

The collections at RBGE represent hundreds of years of exploration as men and women risked their lives to build the scientific resource that we now hold. Some of these people never returned from their travels, but their collections serve as a continual reminder of their commitment and sacrifice. This selection dips a toe into the history of botanical exploration as missionaries, doctors, soldiers and plant hunters collected, painted and photographed the flora of the world.

Chaptalia integerrima

Charles Darwin (1809–1882) was only 22 years of age when he embarked on one of the most famous voyages of all time. He set out from Portsmouth Sound on 27 December 1831 aboard HMS *Beagle*, captained by Robert FitzRoy. During his circumnavigation of the world, Darwin collected over 1,400 plant specimens which he sent to his mentor the Rev. John Stevens Henslow, Professor of Botany at the University of Cambridge. Henslow gave some of these specimens to George Arnott Walker Arnott (1799–1868) and they were deposited at RBGE as part of a permanent loan from the University of Glasgow in 1965. RBGE's is one of only seven herbaria worldwide that have material collected by Darwin; the 75 specimens we hold represent the second largest collection outside of Cambridge. *Chaptalia integerrima* is a common grassland species native to tropical South America where it has become widely naturalised. It was collected by Darwin on the coast of Argentina, close to Bahía Blanca, during the second year of his voyage. **MG**

Collected by Charles Darwin
Voyage of the Beagle 1831–1836

Chaptalia integerrima (Vell.)
Burkart
Det. D.M. Porter 17 June 1981

A Letter from Charles Darwin

In addition to the collection of his specimens in the Herbarium (see p. 50) Charles Darwin also makes an appearance in the RBGE Archive as one of the many correspondents of John Hutton Balfour, Regius Keeper between 1845 and 1879. Darwin wrote to Balfour in 1866, asking him to recommend any Edinburgh nurserymen who could supply him with coloured cowslips or *Primula* species so that he could conduct experiments by cross-pollinating the plants. The Balfour correspondence fills 12 volumes and illustrates the scientific networks which existed during his time at RBGE. **LP**

ROYAL BOTANIC GARDEN
EDINBURGH
E00615974

Angel's Trumpet

Brugmansia suaveolens

This specimen came from the herbarium of George Arnott Walker Arnott, Professor of Botany at the University of Glasgow from 1845 to 1868. It is one of an estimated 100,000 specimens that once formed the core of the University of Glasgow Herbarium and was incorporated into the RBGE Herbarium in 1965. This herbarium sheet bears two different collections of *Brugmansia suaveolens*, a practice that was frequently followed in the 19th century but is generally no longer encouraged in the preparation of modern-day collections. John Gillies and John Tweedie were early 19th-century Scottish botanists and explorers drawn to a new life in southern South America, where they discovered many species new to science. *B. suaveolens* is a very handsome flowering shrub endemic to the coastal forests of south-east Brazil but widely cultivated. Due to the presence of tropane alkaloid, every part of this plant is poisonous, something for which the family Solanaceae is well known. Even with these poisonous properties, many South American cultures use this species in religious ceremonies because of its hallucinogenic effects. **MG**

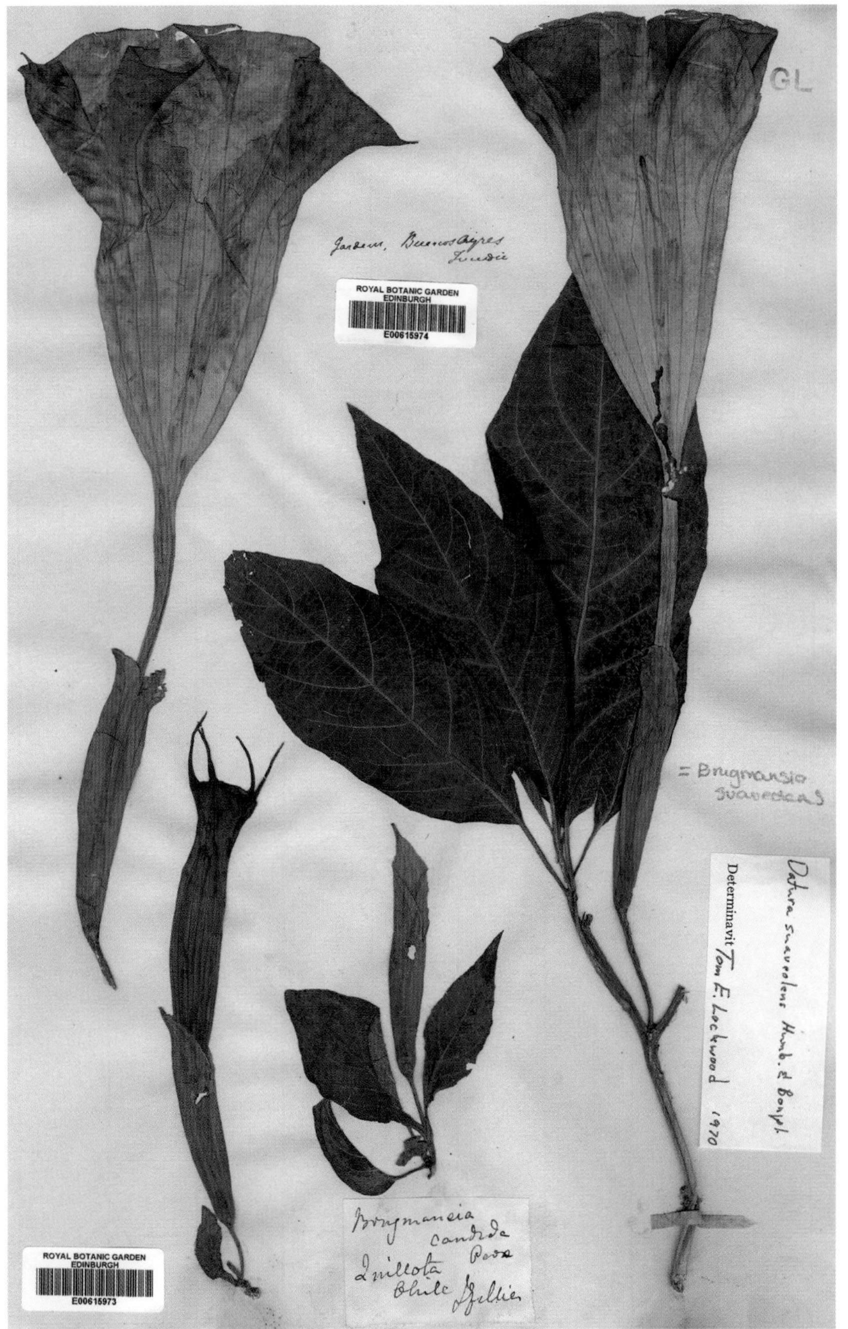

No 3 Amomam Cardamomum. Wild Ginger
In most low shaded lands this Plant is frequent. it grows to five or Six feet high. The flower
Spike is put forth from the Roots and rises to two feet, it is adorned with flowers yellow and red
and succeeded by Berries at first red then Black: containing seeds of the taste of Cardamums
The Roots are large softer than the true Ginger tho often substituted in its stead by the makers
of Sweetmeats in these parts.
Grass
No 47 millet Grass
No 46

Hortus Siccus Jamaicensis

This magnificent three-volume set is the grandest of several versions made by the surgeon and physician William Wright (1735–1819) between 1782 and 1785, during the second of his two periods in Jamaica. John Hope had a two-volume version. Wright, who "suffered no moral anxiety about slavery" and was a slave owner himself, was interested in the diseases suffered by slaves (especially 'yaws') and the local economic and medicinal uses of plants. These uses are carefully documented in manuscript on the opposite side from the mounted specimens. This set was presented to the museum of the Society of Antiquaries of Scotland and dedicated to its founder, David Steuart Erskine, 11th Earl of Buchan. Behind it lies a fascinating story of rivalry in the setting up of this Whiggish society (its museum curated by Hope's pupil William Smellie) and a Tory alliance of the Royal Society of Edinburgh with the University of Edinburgh (and its museum). In view of this it is ironic that these volumes should have ended up in the collection of the University of Edinburgh and thence to RBGE. **HN**

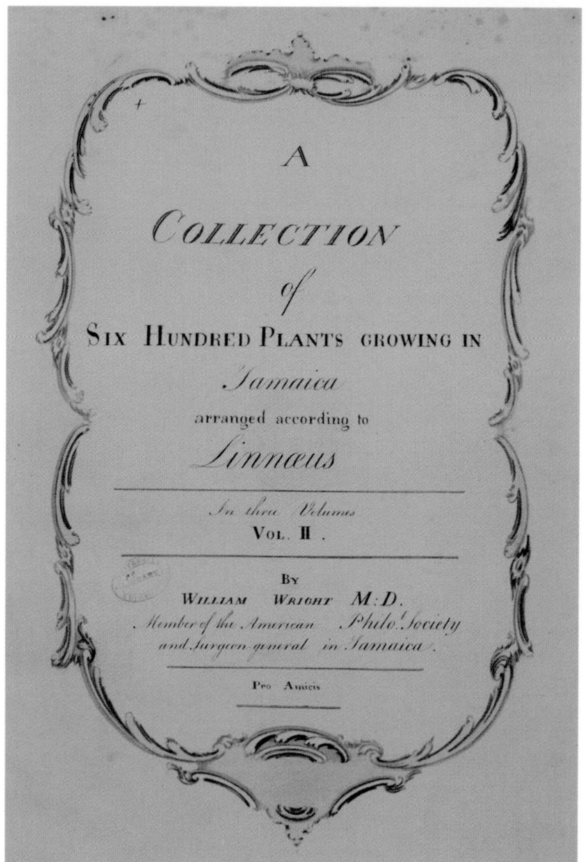

A

COLLECTION

of

SIX HUNDRED PLANTS GROWING IN

Jamaica

arranged according to

Linnæus

In three Volumes

VOL. II.

BY

WILLIAM WRIGHT M:D.

Member of the American Philo! Society
and Surgeon general in Jamaica.

Pro Amicis

David Douglas' Telescope

Scots-born David Douglas (1799–1834) is known for the introduction to science and horticulture of trees such as the sugar pine, ponderosa pine, Sitka spruce and the Douglas fir – named in his honour. Douglas met an untimely death, his body being discovered in a bull pit trap on Mauna Kea in the Sandwich Islands – now Hawaii. He was at first believed to have fallen into the pit accidentally, but later investigations suggest that he was murdered by his guide, a former convict from Botany Bay. This telescope was given to Douglas by William Wells of Redleaf estate in Kent and was apparently found in the bull pit containing his body. It was sent to his brother George and eventually became the property of F.R.S. Balfour of Dawyck, who, in turn, donated it to RBGE. **LP**

East Maui Silversword

Argyroxiphium sandwicense

The plant collector David Douglas came across this remarkable plant of the family Compositae (daisies) when he was nearing the summit of Mauna Kea. This was one of the last collections he made before he died on the same mountain a few months later (see p. 56). The bravery of collectors like Douglas, who risked their lives collecting plants, is a major reason that we now have these extraordinary scientific resources. This species is endemic to Hawaii and the tall flowering columns can reach 3 metres. It is monocarpic, that is long-lived but flowering just once in its life, and is currently listed as critically endangered; there may be only 40 wild individuals left. **EH**

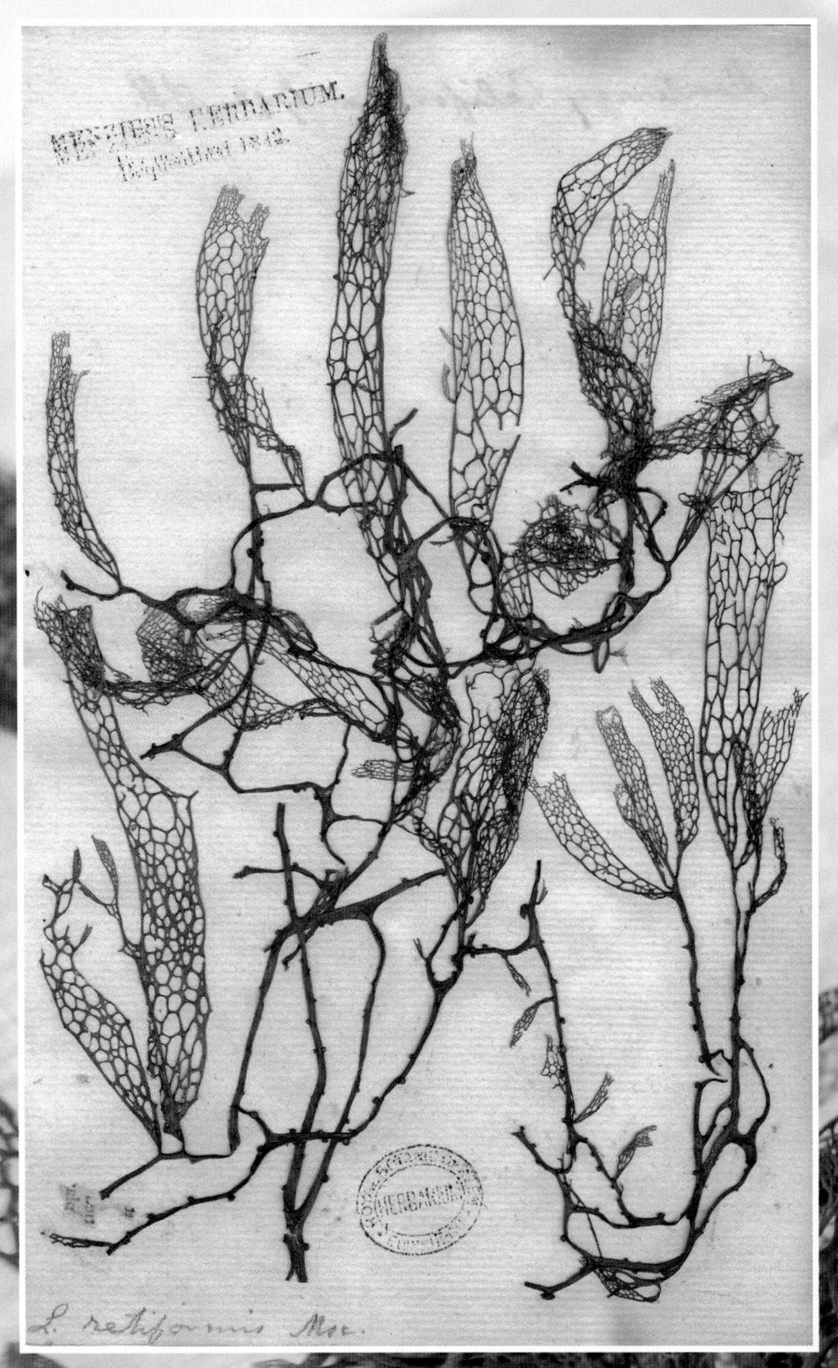

Lace Lichen

Ramalina menziesii

Ramalina menziesii drapes the trees of coastal California, particularly where coastal fogs roll in. It was one of the first collections from North America, made in the late 18th century by Archibald Menzies (1754–1842), the tireless plant hunter and ship's surgeon who was on an expedition to circumnavigate the globe. Menzies is recognised for many important botanical discoveries, with nearly a hundred species named after him. His lichen collections are probably of equal importance and form the foundation of North American lichenology. He distributed his collections widely among active researchers in the field. The fact that this specimen was mounted upside-down on a sheet, as if it had been growing upward like a seaweed, illustrates the confusion that surrounded the species. Though Menzies and others recognised its correct taxonomic position with the lichens, it was in fact described as an alga before its classification with the lichens was widely known. **RY**

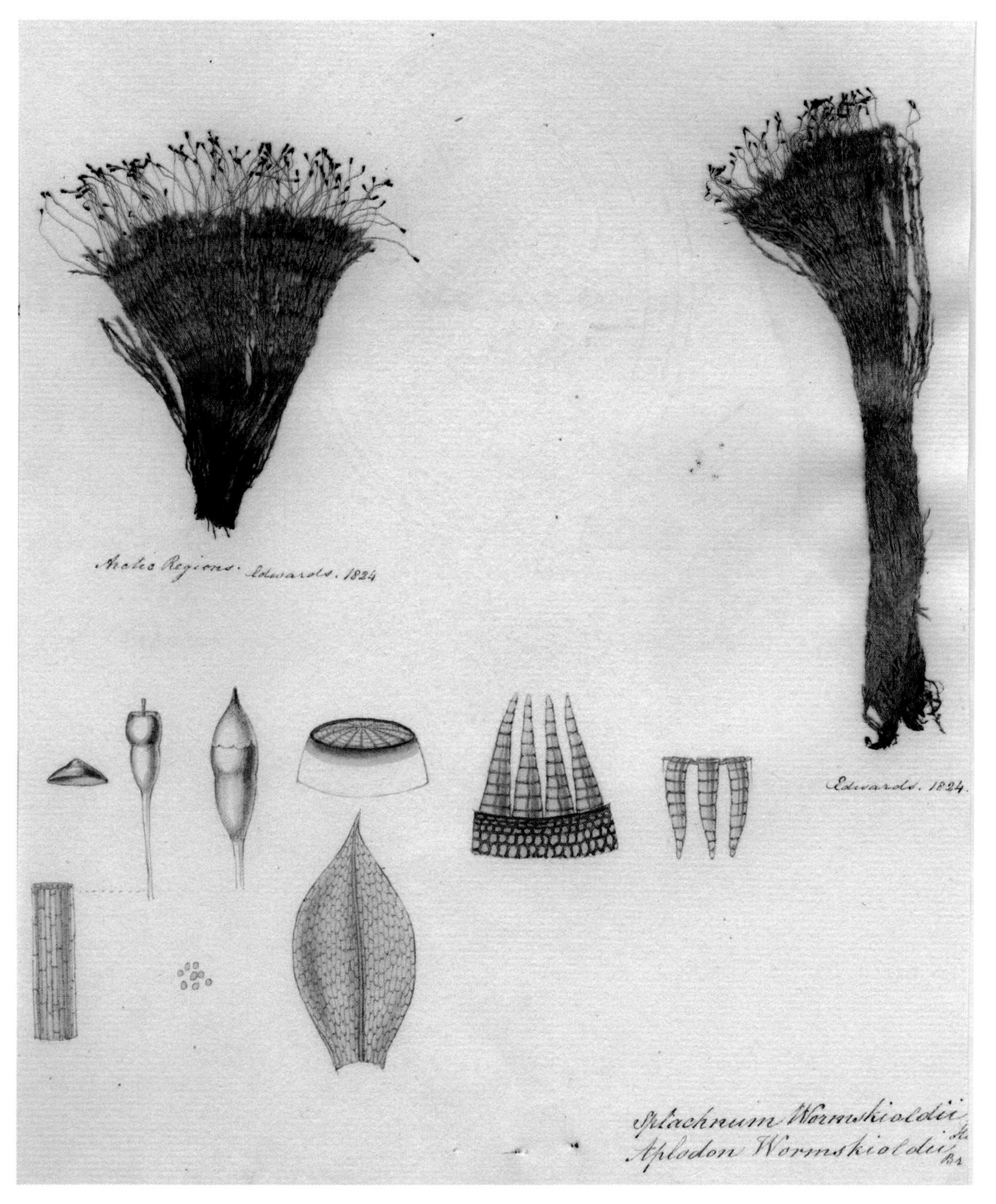

Carrion Moss

Aplodon wormskioldii

John Edwards was a ship's surgeon on Sir William Edward Parry's three expeditions to search for the Northwest Passage between 1819 and 1825. This specimen of the moss *Aplodon wormskioldii* was collected during the third and last attempt by Parry to find a navigable route between the North Atlantic and Pacific Oceans. It was sent to Robert Kaye Greville, who studied the structure of the various parts and produced these exquisite and scientifically precise watercolour sketches. This is a species which survives well on the high nitrogen matter of lemming dung and is also frequently found on the bodies of dead lemmings, giving rise to its common name of carrion moss. **EH**

Wild Lupine

Lupinus perennis

Sadly the Arctic expeditions of Sir John Franklin are now more renowned for cannibalism than for the botanical, zoological, geological and meteorological discoveries that were made. Taking place at about the same time as the first two Arctic expeditions of Sir Edward Parry, the first two Franklin expeditions included the talented naturalist John Richardson (1787–1865), who was largely responsible for the collections that were made on the voyages. He was one of the few to survive the tragic first expedition to search for the Northwest Passage. While there appears to be proof of cannibalism during the last days of the third and final expedition, from which no survivors returned, there were also rumours of it during the first expedition by a man who was later accused of murder and shot by Richardson. This little lupine, collected on the shores of the Arctic Sea, is one of the plants Richardson collected. **EH**

Torbay
GREVILLE

Sphaerococcus coronopifolius

As one of the most respected and renowned experts on algae of her day, Amelia Warren Griffiths (1768–1858) of Torquay appears to have possessed both intellect and beauty. "Mrs Griffiths, than whom no one is able to form a better judgement, or whose opinion is entitled to greater weight," in the words of the famous 19th-century botanist George Bentham, was described by William Jackson Hooker as an "ornament of her sex". Living in Torquay, she sent many specimens to Robert Kaye Greville who later published the *Algae Britannicae*. This species of red alga shows the delicate structure and beautiful colour which made it so popular among Victorian naturalists. The specimen would have been mounted by floating it onto the sheet to ensure that the most intricate tips of the branches were perfectly presented. **EH**

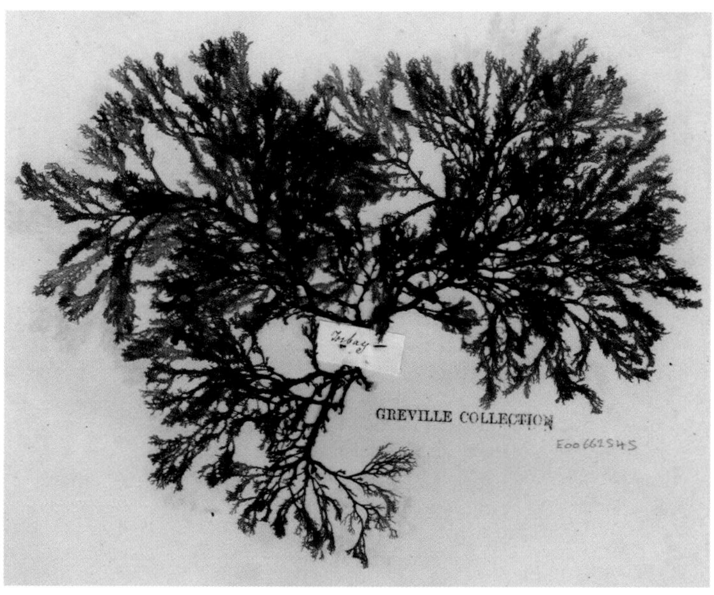

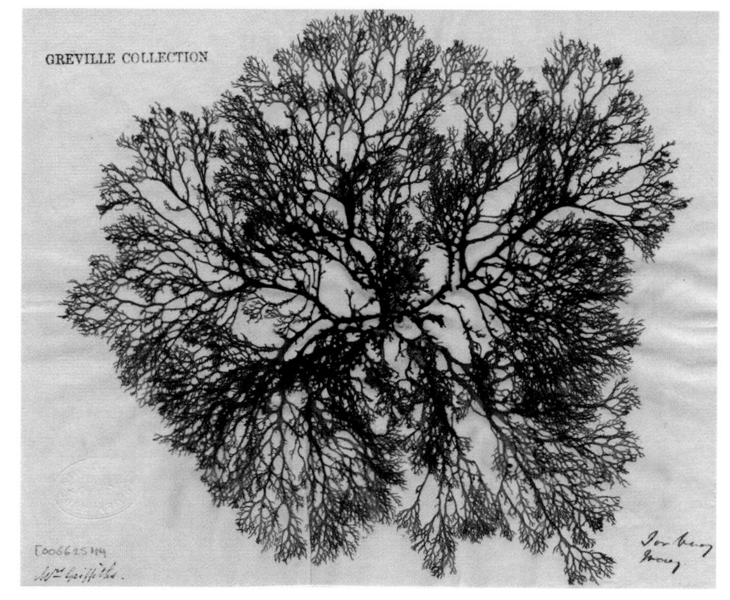

Edelweiss

Leontopodium nivale subsp. *alpinum*

The RBGE Herbarium has grown through collection by staff members, donations by individuals and other organisations, and sometimes through purchase. The collections from Europe ('Area 1') are extensive, and one of the largest elements was purchased in 1912 from the Austrian botanist Ignaz Dörfler (1866–1950), during Isaac Bayley Balfour's tenure as Regius Keeper. Dörfler obtained material from many collectors such as this edelweiss (collected by Markus, Freiherr von Jabornegg in the Carinthian Alps of southern Austria), which he curated as a collection called 'Herbarium Normale' that he distributed by sale. A sheet from another collection notes that edelweiss is "met with only on the highest mountains of Tyrol & Bavaria. It always grows in a spot to be reached only with the greatest peril". This extreme habitat is confirmed, uniquely, by photographs of the collector in action on another specimen showing "Heli Tränker expert climber" collecting by abseiling, for Isobel Case. Case was curator of the University of Glasgow Herbarium, the foreign part of which was given to RBGE in 1965. **HN**

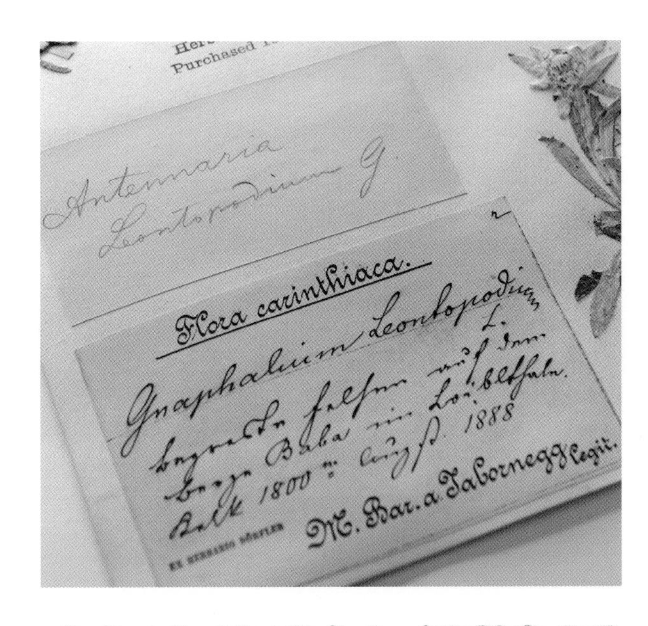
Antennaria
Leontopodium G.
Flora carinthiaca.
Gnaphalium Leontopodium L.
1888
M. Bar. a Jabornegg legit.

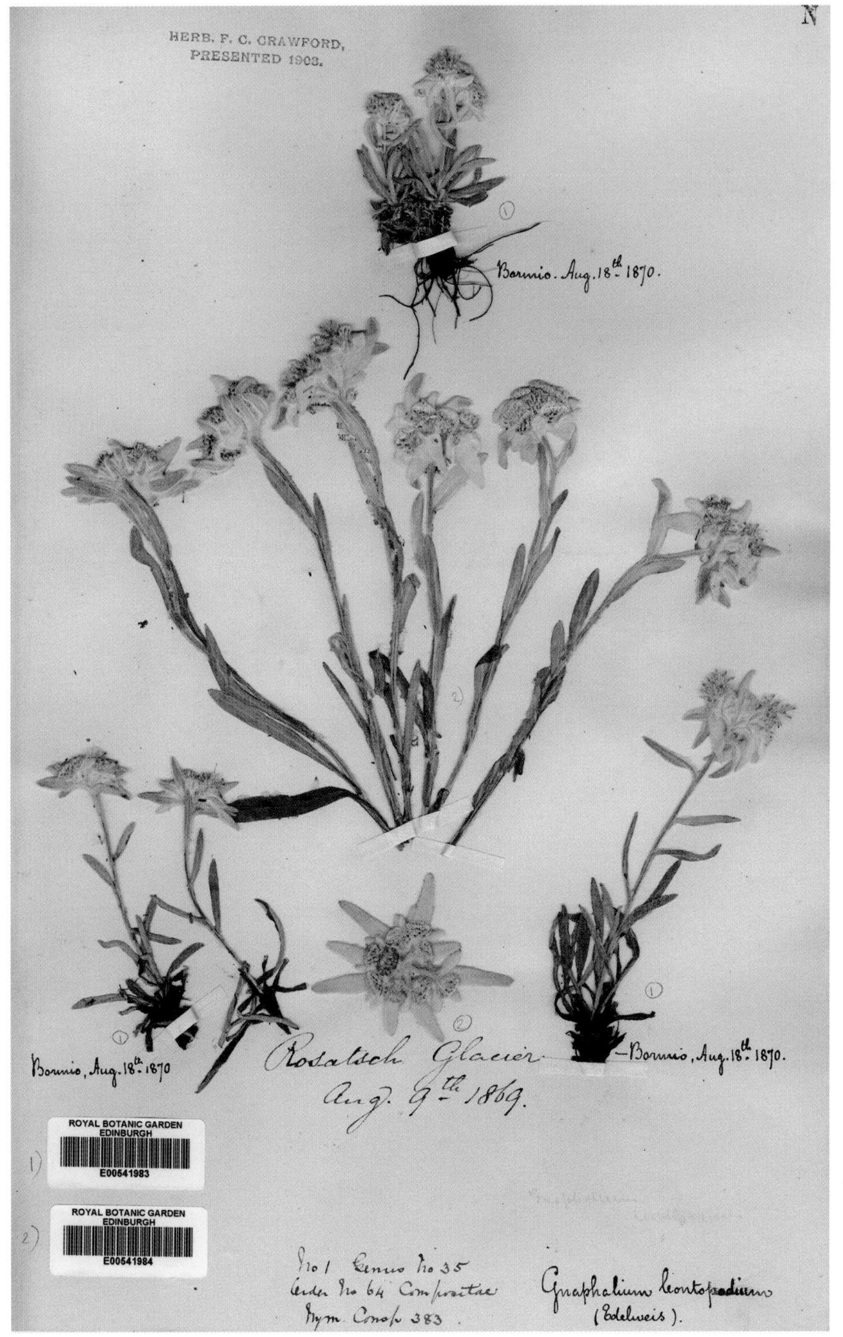
HERB. F. C. CRAWFORD,
PRESENTED 1903.
Bormio. Aug. 18th 1870.
Bormio, Aug. 18th 1870
Rosatsch Glacier
Aug. 9th 1869.
Bormio, Aug. 18th 1870.
ROYAL BOTANIC GARDEN
EDINBURGH
E00541983
ROYAL BOTANIC GARDEN
EDINBURGH
E00541984
No 1 Genus No 35
Order No 64 Compositae
Nym. Consp. 383.
Gnaphalium leontopodium
(Edelweis).

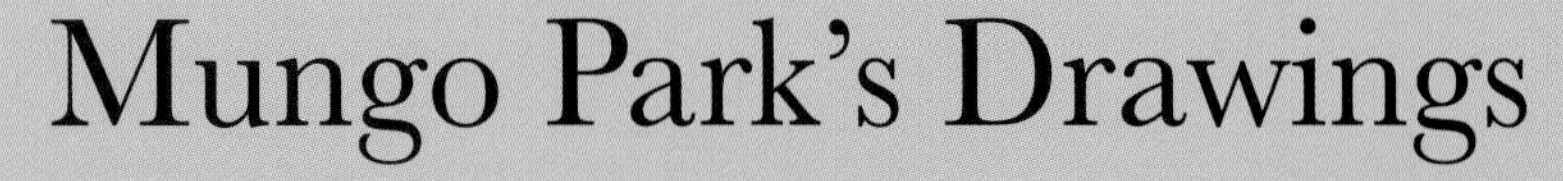

Mungo Park's Drawings

Mungo Park (1771–1806) trained as a surgeon in Selkirk and studied medicine in Edinburgh. His eldest sister married fellow Borderer James Dickson, a London-based nursery and seedsman and founder member of the Royal Horticultural Society of London. In 1792, Dickson and Park made a botanical tour of Scotland, visiting Ben Lawers and Ben Nevis amongst other localities. Park had a self-confessed "general passion for travelling" which saw him sail to Sumatra as a surgeon on HMS *Worcester* and lead two expeditions to West Africa to explore the River Niger. He first set foot on African soil in June 1795 and travelled up the Gambia to Pisania, where he stayed for six months fact-finding and learning local languages. During this time he contracted a severe fever (possibly malaria) from which he recovered. He explained that to pass some of his convalescence, "I amused myself with drawing plants &c. in my chamber." The sketch book held at RBGE contains 53 of Park's drawings. Drawings of a turtle and a vibrant yellow bat indicate that he did not only draw plants. **GH**

Sketches by Mungo Park, during his visit to Dr. Laidley; from the end of June to the second of December 1795. He was ill with fever in August and relapsed on the tenth of September. As several of the sketches are dated by Park, it appears that they must have been made, while residing with Dr. Laidley at his house at Pisania on the Gambia (CMG)

Fissidens parkii

Mungo Park's 1795 expedition made him the first British explorer to journey into West Africa and his sketches are possibly the first botanical drawings from the area. This moss could be considered to have saved Park's life when he first became ill with fever, as he states in a letter, "I considered my fate as certain and that I had no alternative but to lie down and perish. At this moment, painful as my reflections were, the outstanding beauty of a small moss (*Dicranum bryoides*) in fructification, caught my eye. I mention this to show from what trifling circumstances the mind will sometimes divine consolation, for though the whole plant was not larger than the top of my fingers I could not contemplate the delicate information of the moss, leaves and capsules, without admiration." This event is illustrated in the frontispiece of the 1824 book *The Wonders of the Vegetable Kingdom Displayed* by Mary Roberts. It is appropriate that this moss was named after Park when it was found to be a new species. Sadly, Park's life ended on his second expedition to Africa in 1806 while he was charting the River Niger. His wife, crushed by grief, never accepted his death, and one of his sons died in an expedition to find his father. **EH**

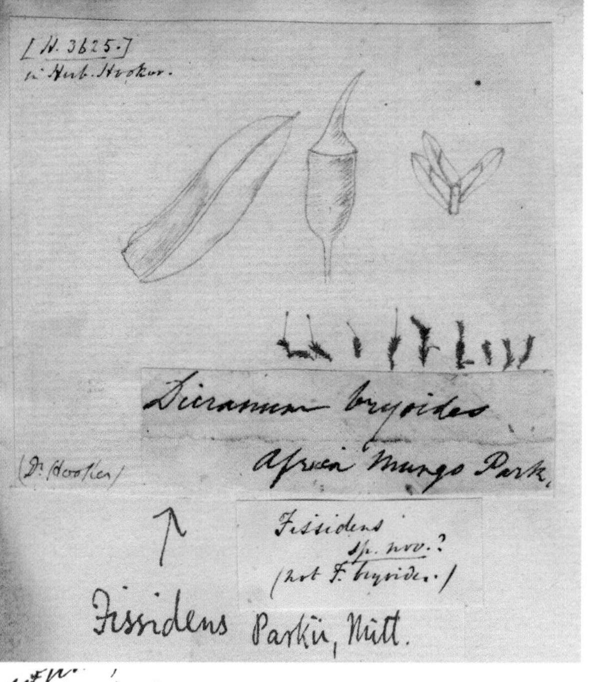

"Whichever way I turned, I saw myself in the midst of a vast wilderness, in the depth of the rainy season, naked and alone; surrounded by savage animals; and by men still more savage. I wa 500 miles from any European settlement. All these circumstances crowded at once on m recollection and I confess that my spirits b to fail me. I considered my fate as certain that I had no alternative but to lie down perish. At this moment, painful as my tions were, the extraordinary beauty of Dicranum bryoides) in fructifica to shew from wh mind will sometimes plant was

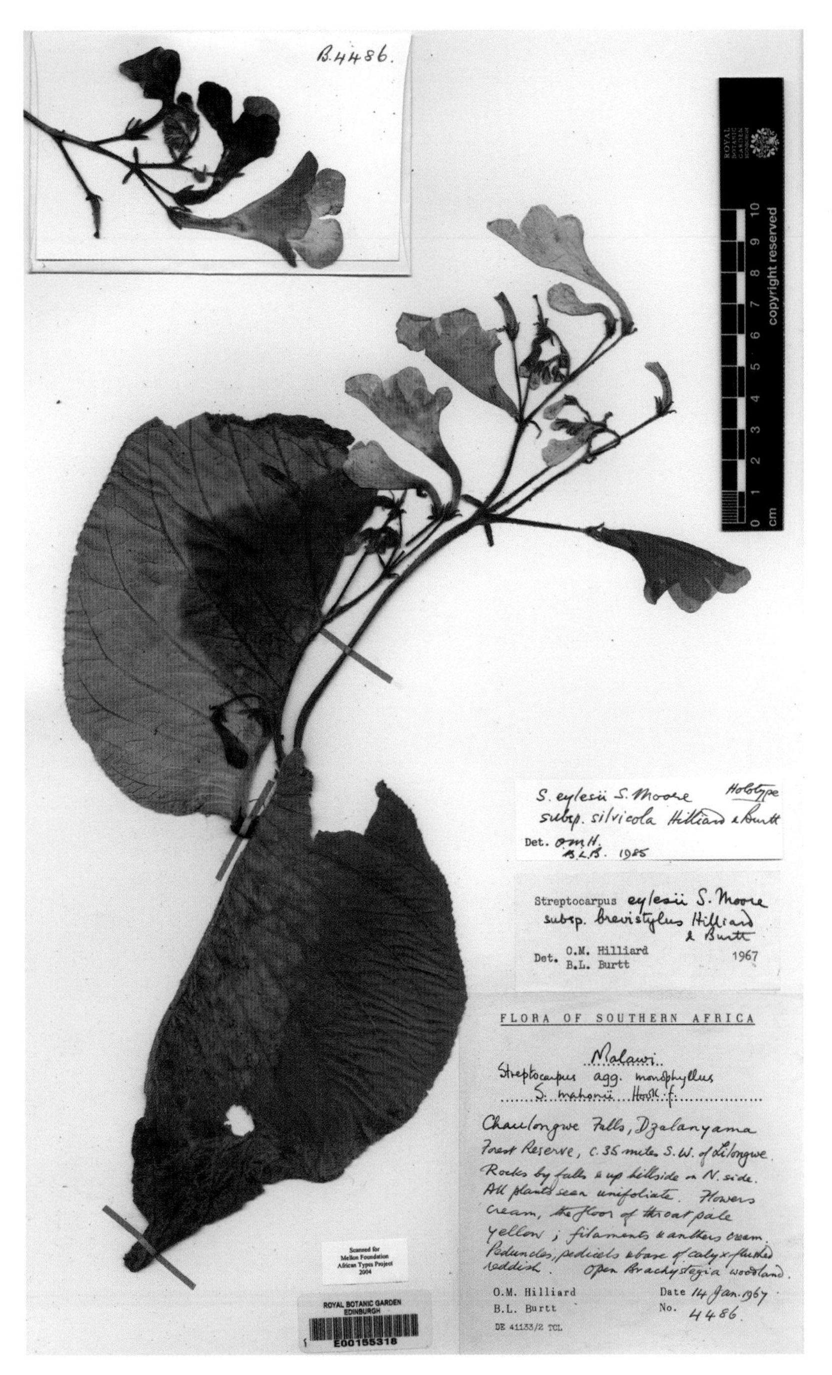

Streptocarpus eylesii subsp. *silvicola*

Between them Olive Hilliard (1926–) and Brian Laurence Burtt (1913–2008) collected over 22,000 specimens, of which 5,000 were collected during their long and successful collaboration. This began in 1964, the year that the new Herbarium and Library were opened at RBGE, and Burtt played a key role in the design of the new building. Together the pair described several hundred new species, many of which were to become popular garden and house plants. Although they gained expertise across the plant kingdom, it was to the Gesneriaceae that they devoted much of their later careers. The genus *Streptocarpus*, commonly known as the Cape primrose, was to become a particular passion. This specimen, collected in Malawi in 1967, demonstrates the detailed collection notes which were a feature of their work. Four years later, they published their seminal book on the genus *Streptocarpus: an African plant study*. **EH**

Malawi.
Streptocarpus agg. monophyllus
S. mahonii Hook. f.

Chaulongwe Falls, Dzalanyama Forest Reserve, c. 35 miles S.W. of Lilongwe. Rocks by falls & up hillside on N. side. All plants seen unifoliate. Flowers cream, the floor of throat pale yellow; filaments & anthers cream. Peduncles, pedicels & base of calyx flushed reddish. Open Brachystegia woodland.

O.M. Hilliard Date 14 Jan. 1967
B.L. Burtt No. 4486.
DE 41133/2 TCL

Davallia species

There is very little information for this specimen collected during David Livingstone's (1813–1873) six-year exploration of the River Zambesi, which began in 1858. It was a tragically unsuccessful expedition during which Livingstone's wife died of malaria. The physician John Kirk was one of the expedition and he was responsible for the large numbers of botanical specimens that were collected and sent back to scientific institutes in Britain, of which a large number proved to be new species. This specimen is one of very few known to be held in the Herbarium at RBGE from this expedition. It has not yet been identified to species level. Over the last 300 years of taxonomic work at RBGE a huge amount has been done to identify specimens in order to ensure that we have a record of the plant species growing around the world. However, there is still a daunting amount of work to be done in identifying specimens that are currently held in herbarium cabinets. **EH**

Illustrations of Soqotran Plants

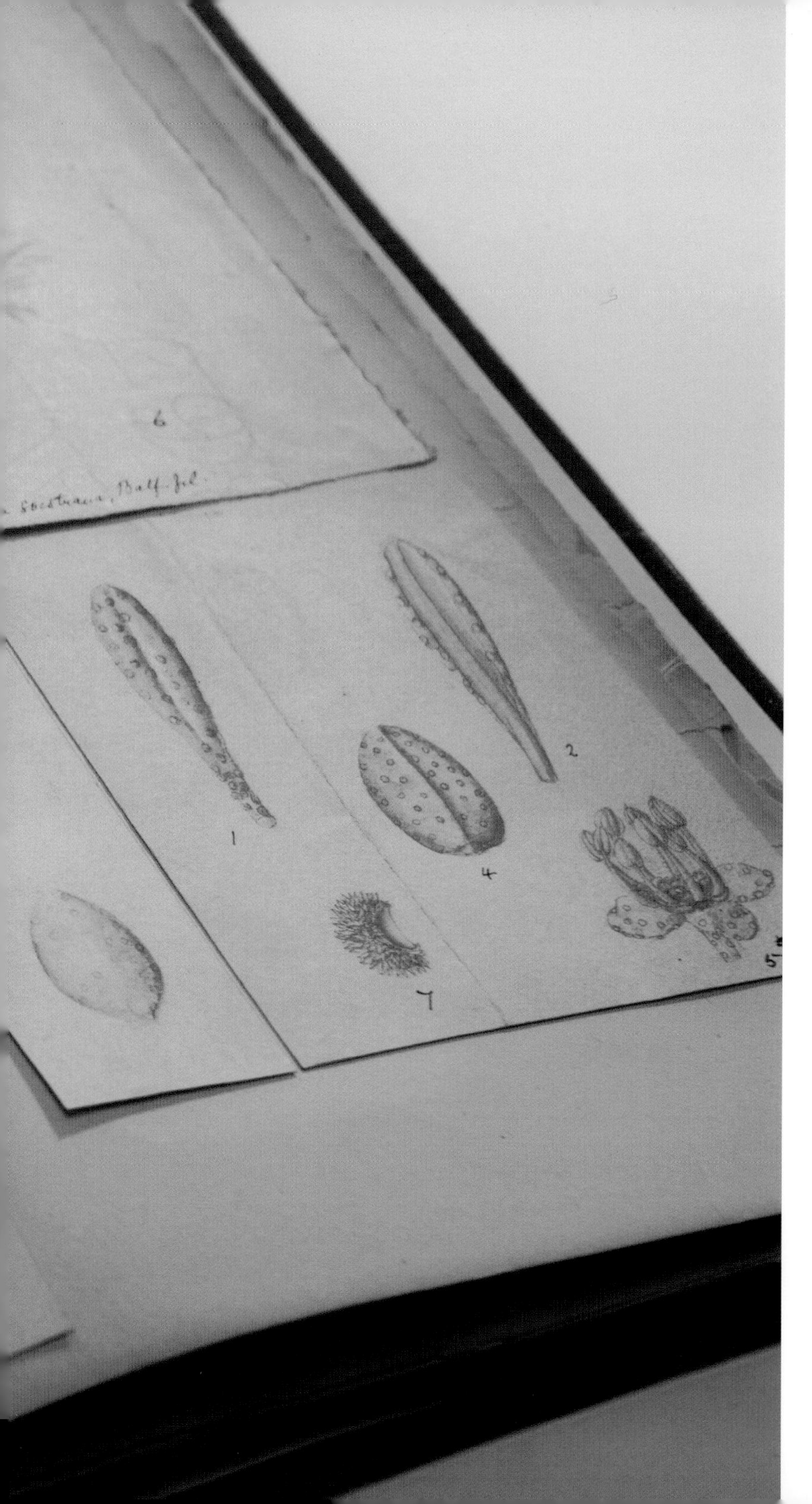

Son of Regius Keeper John Hutton Balfour, Isaac Bayley Balfour followed his father into medicine, botany and eventually the post of Regius Keeper in 1888. Prior to this date his considerable achievements included taking part in an expedition to Rodriguez, becoming the first student to obtain the University of Edinburgh's Doctor of Science degree and holding the posts of Professor of Botany at the Universities of both Glasgow and Oxford. He also travelled to the island of Soqotra during the winter of 1879–1880, under the auspices of the Royal Society and the British Association, to study its geology and botany over just seven weeks. He collected a considerable volume of information and plant specimens, including nearly 200 new species of flowering plants and almost 70 lichens. Some of the beautiful pencil illustrations produced as part of his study are shown here, drawn by Harriet Thiselton-Dyer (1854–1945), daughter of former Kew director Joseph Dalton Hooker (1817–1911) and wife of William Thiselton-Dyer, Hooker's successor. **LP**

ROYAL BOTANIC GARDEN EDINBURGH
52/80
27
ROYAL BOTANIC GARDEN
EDINBURGH
E00188956
Dracaena cinnabari Balf.f.
Det.: J.J. Bos
B.C.S. 11.
SOCOTRA PLANTS.
COLLECTED BY DR. BAYLEY BALFOUR.
PRESENTED, 1888.

Dragon's Blood Tree

Dracaena cinnabari

Dragon's blood resin has been known as a commodity since Classical times. One of the sources of this resin is the dragon's blood tree, *Dracaena cinnabari*, which grows in Soqotra. The 1880 expedition to the island by Isaac Bayley Balfour discovered the tree growing there to be a distinct species. It was restricted to the higher ground, above 1,000 feet (approximately 300 metres). He reported that the resin was harvested from the trees immediately after the rains by using goat skins to catch the sap from a cut made with a knife. In his publication *The Botany of Socotra* almost 300 new species were described. **EH/SK**

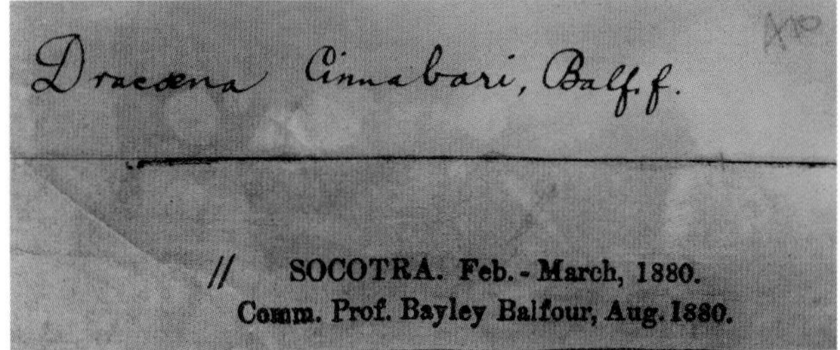

Euphorbia davisii

Peter Hadland Davis (1918–1992) was one of the most prolific collectors and notable plant taxonomists of the 20th century. He spent his professional life at RBGE, employed by the University of Edinburgh, where he held a personal chair in Taxonomic Botany. Between 1938 and 1986 he collected some 71,163 herbarium specimens – from Malaysia to Brazil, via the Crimea, Morocco and Madeira, but especially in Turkey, having become interested in the Eastern Mediterranean flora while undertaking security work during the Second World War. He also collected contemporary paintings and Wemyss pottery. Davis's great work, in addition to teaching, and the authorship (with V.H. Heywood) of the influential 1963 *Principles of Angiosperm Taxonomy*, was the *Flora of Turkey*, started in 1959 and published, with the help of a dedicated team, in nine volumes between 1965 and 1985, covering almost 8,500 species. This included many described as new to science, such as this spurge, named after Davis by Mohammed Salar Khan, a doctoral student from what was then East Pakistan. **HN**

Common Poppy

Papaver rhoeas

It was felt that a book published in the centenary year of the First World War should contain an image of a red poppy. In the Herbarium there are no specimens of poppies collected in Flanders, but this one is a reminder of an earlier European war – the Crimean. The specimen was collected by John Kirk (1832–1922) who, after training in medicine at Edinburgh, which included the study of botany under John Hutton Balfour at RBGE, went to work at the hospital at Renkioi on the Asian side of the Straits of Dardanelles. This hospital was a prefabricated, wooden structure, designed by Isambard Kingdom Brunel to have better sanitation, and to be erected in a more salubrious situation, than Florence Nightingale's beleaguered hospital at Scutari. In his free time Kirk went bustard shooting and plant collecting, resulting in RBGE's earliest Turkish collections. Kirk later joined Livingstone's second Zambesi expedition (see p. 69), on which he made pioneering photographs and botanical collections; he later achieved fame (and a knighthood) through his anti-slavery work in Africa, especially in Zanzibar. **HN**

Sacred Lotus

Nelumbo nucifera

With the transfer to RBGE of the Cleghorn Memorial Library and a vast collection of illustrations, both prints and original drawings, from the Royal Scottish Museum in 1941, Hugh Cleghorn (1820–1895) posthumously became one of the greatest benefactors to the RBGE Library and Archive. The material joined his herbarium which had been presented to the Garden by his nephew in 1896. Cleghorn trained as a surgeon at Edinburgh and studied botany under Robert Graham in 1836 and 1837; he then joined the East India Company. During his first period in India he spent much time in Mysore, in the present-day Indian state of Karnataka, based at the town of Shimoga, where he acted as surgeon, magistrate and supervisor of jails. Between July 1845 and July 1847 Cleghorn employed a 'Marathi' artist to paint a different species each day, and these drawings form a fascinating visual botanical diary. This beautiful image of the sacred lotus comes from this collection. **HN**

The Grasses of Assam

Samuel Edward Peal (1834–1897), who trained as an artist, went to Assam in north-east India as a tea planter in 1862 where he made a detailed illustrated study of the grasses of his tea estate near Sibsagar on the Brahmaputra floodplain. The resulting elephant folio volume includes an introductory section on the structure of grasses, followed by depictions in delicate watercolour of individual species with accompanying notes and local names. The surgeon James Murray Foster (1836–1879), who worked for the Assam Company based at Nazira, decided to make a facsimile of Peal's work "to pass away the weary hours of the wet sultry afternoons during the rainy season" and "to secure a duplicate … in case of an accident to the original". Prophetic words, as Peal's bungalow and all his research papers were later destroyed by fire. This is a recent donation to the RBGE Library by the Cann family, descendants of James Murray Foster, made because of our interest in the flora of north-east India. **HN**

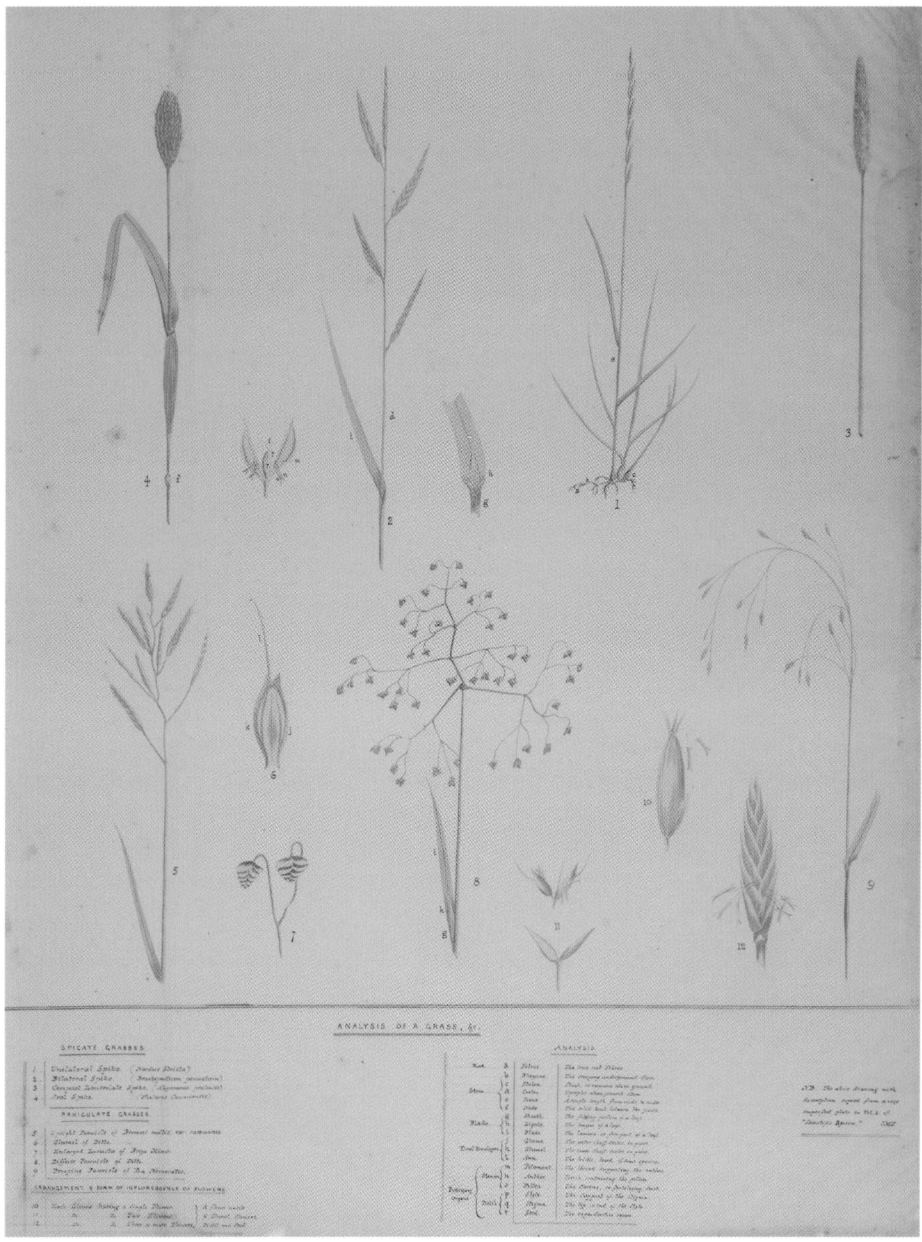

Flamboyant Tree

Delonix regia

This is one of a collection of 170 drawings made for Alexander Gibson (1800–1867) in the botanical garden of Dapuri near Poona. Gibson was an East India Company surgeon, born at Laurencekirk, who studied medicine at Edinburgh, including botany under Daniel Rutherford at RBGE. Most of his professional life was spent in the Bombay Presidency in western India, of which, in 1847, he was appointed its first Conservator of Forests. Earlier he had been appointed to superintend the botanical garden of the Presidency, which was situated in the grounds of the Governor of Bombay's summer residence. The garden existed from 1827 until 1865 and in it Gibson grew both native species (encountered on his forest tours) and exotics such as this spectacular leguminous tree, the 'flamboyant' or 'gul mohur', which is native to Madagascar, but is widely grown as a street tree throughout the tropics. For a period of 26 months in 1847–1848 and 1849–1850 Gibson employed an artist to draw the plants in the garden – he failed to record the artist's name, noting only that he was a Portuguese-Indian. **HN**

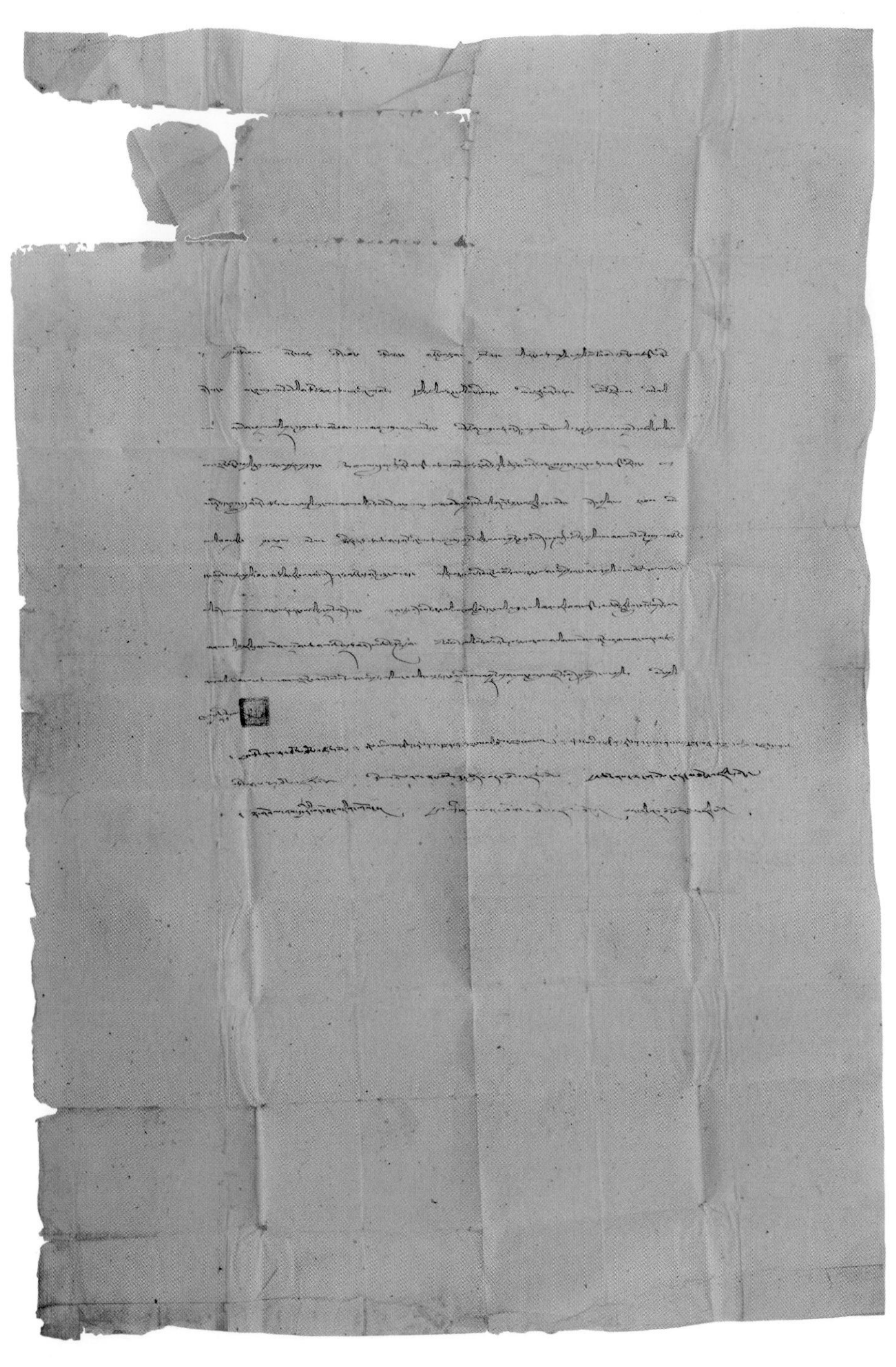

Tibetan Passport – 'lamyig'

This passport, written on paper made of *Daphne* bark, was issued on the "21st day of the 7th Tibetan month of the Fire-Dog year [17 September 1946]" and was negotiated by Hugh Richardson, the last British representative to the Tibetan Government in Lhasa. It allowed the final expedition to Tibet of collectors Frank Ludlow (1885–1972) and George Sherriff (1898–1967), which took place from October 1946 to October 1947. The party consisted of Major and Mrs Sherriff, Frank Ludlow, Colonel Henry Elliot of the Indian Medical Service, a Tibetan headman, two Nepali cooks, four Lepcha plant collectors, and the labradors Joker and Jill. The aim was to explore the little-known province of Pome in south-east Tibet and it resulted in the collection of some 2,822 herbarium specimens and 20,000 packets of seed. The passport instructs local headmen ('dzongpons') to provide transport at a specified rate ("twelve sangs per pony and six sangs per pack animal for a day's journey") and food at reasonable prices, but that the visitors should not stray from the agreed route, shoot animals on sacred sites or "beat or ill-treat the Tibetan subjects". **HN**

Saussurea pinnatiphylla

The collectors Frank Ludlow and George Sherriff made a remarkable series of six expeditions to Bhutan and Tibet between 1933 and 1949. Their close links with the Bhutanese royal family gave them unique access to that reclusive country, and it was a set of their beautifully collected herbarium specimens given to RBGE, with ones made by Roland Edgar Cooper (see p. 81) in 1914 and 1915, that formed the cornerstones on which the *Flora of Bhutan* project was built. This *Flora* was started in 1975 by Andrew Grierson, initially in collaboration with David Long; it was completed in nine parts, covering 5,603 species, in 2001. One of Grierson's major interests was the family Compositae, but he died before his account of the family was finished, and the work was completed by Lawrence Springate. Initially this high-altitude rosette plant was identified as a Chinese species, but it proved to be new to science and was eventually published in 2000, 51 years after its collection on the last of the Ludlow and Sherriff expeditions. **HN**

Roland Edgar Cooper's Photographs

Born in Kingston-upon-Thames, Roland Edgar Cooper (1890–1962) was brought up by his aunt Emma, the wife of William Wright Smith (1875–1956), Regius Keeper of RBGE from 1922 to 1956. In 1907, Smith was appointed to direct the herbarium in the Royal Botanic Garden, Calcutta, and Cooper went with him, travelling and collecting botanical specimens in Sikkim, Nepal, Tibet and Bhutan. In 1910, Cooper and Smith returned to Edinburgh and Cooper took the Horticultural course at RBGE. In 1913, Cooper returned to the Himalayas to collect plants for A.K. Bulley of Ness, near Liverpool. He travelled through Sikkim in 1913, Bhutan in 1914–1915 and the Punjab in 1916. In 1921, after the First World War, during which Cooper served in the Indian Army, he was appointed Superintendent of the Botanic Garden at Maymyo in the Shan Hills of Burma. In 1930, he joined the staff at RBGE in the post of Garden Curator's Assistant, moving on to the role of Curator (Head Gardener) in 1934, a post that he held until his retirement in 1950. These images are from albums of photographs that Cooper took during his travels in the Himalayas and Sikkim, probably during his 1913 collecting trip. **LM**

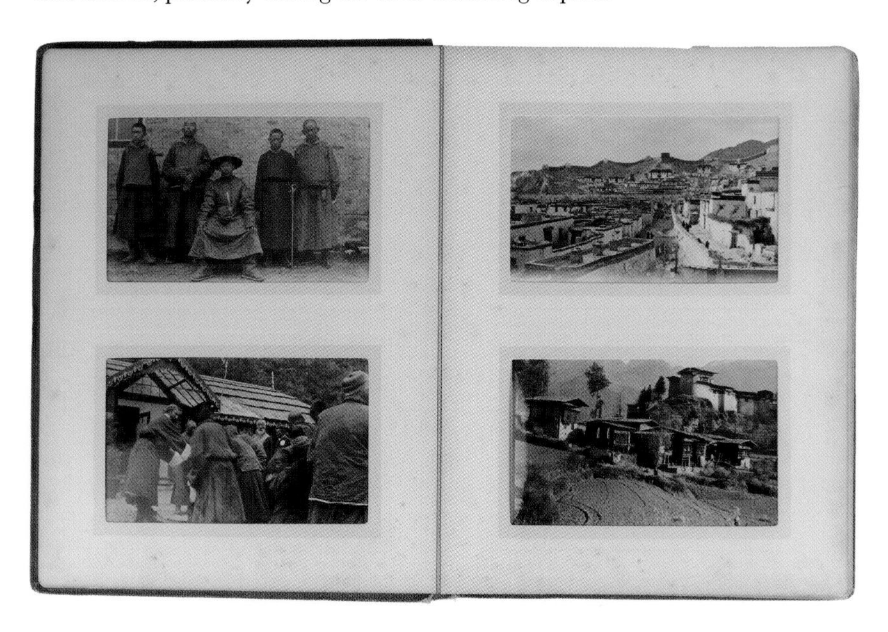

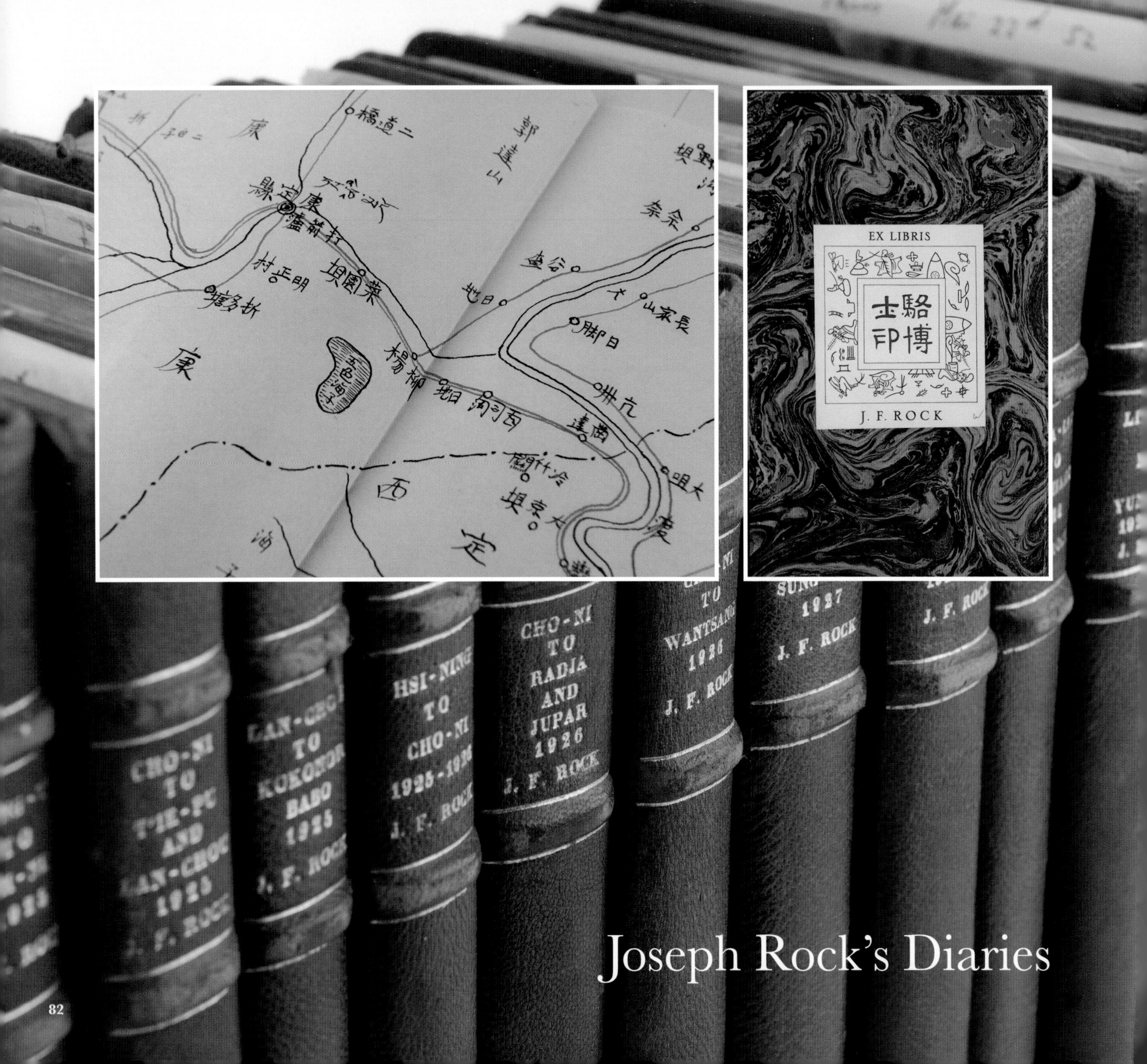

Joseph Rock's Diaries

Joseph Francis Charles Rock (1884–1962), Austrian born but a naturalised American, collected plants in and around Yunnan and Szechuan provinces in south-west China between 1920 and 1949, but he was much more than a plant collector. An eccentric character, he seemed to revel in travelling around China like a prince, followed by a large entourage carrying his home comforts. Behind this was an alter-ego, a quiet loner, full of self-doubt and criticism – a fascinating paradox. Evidence of this, and his descriptions of the people and places he travelled amongst, is revealed in his diaries which were bequeathed to RBGE. Rock was especially interested in the different cultures he came across, making a particular study of the Naxi people who lived in and around Lijiang in north Yunnan and producing the first dictionary of their pictographic script, still used today, and much of which features in the diaries – Rock even used the characters in his bookplate. **LP**

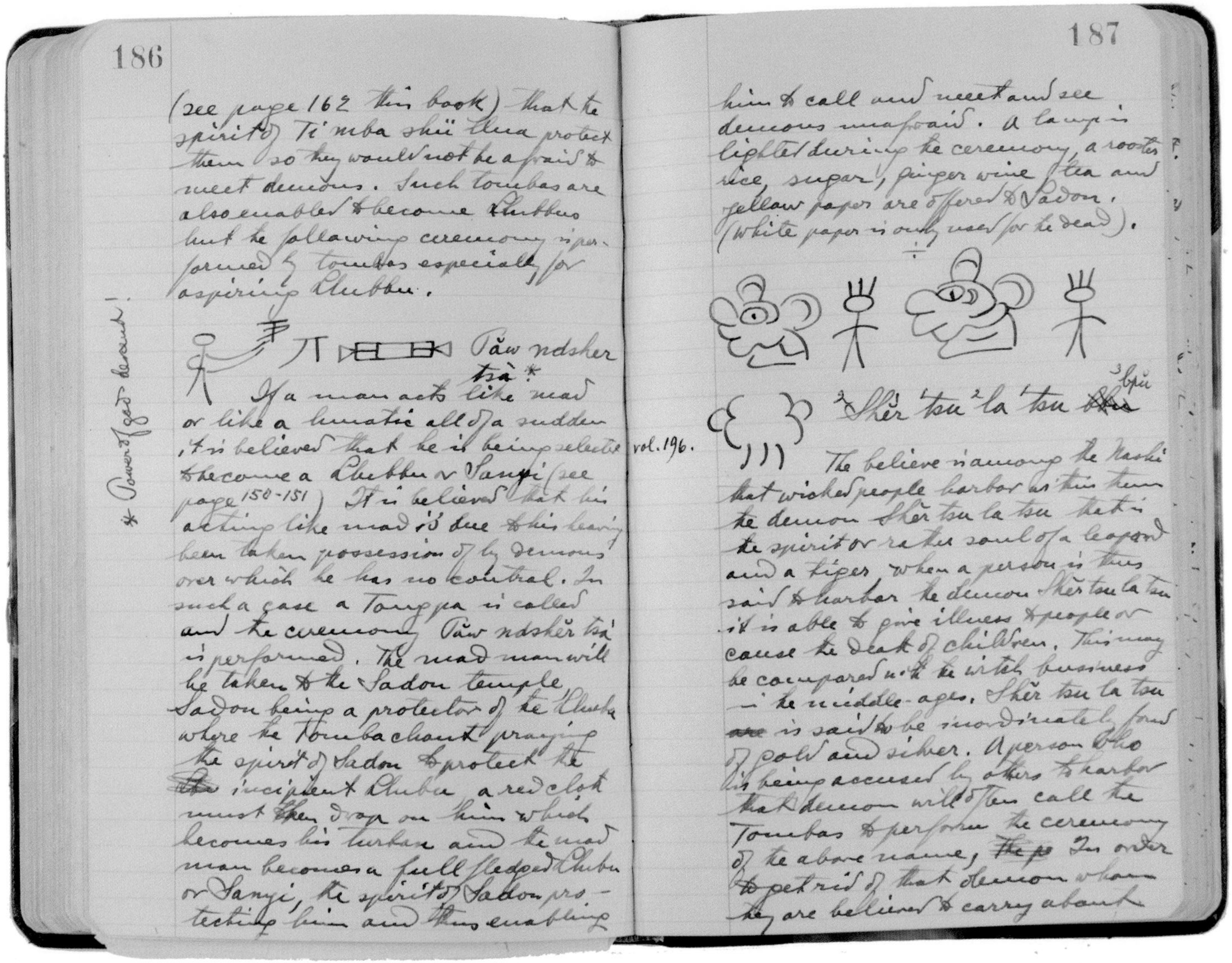
186

(see page 162 this book) that the spirit of Ti'mba shü llua protect them so they would not be afraid to meet demons. Such tombas are also enabled to become Llubbus but the following ceremony is performed by tombas especially for aspiring Llubbu.

Päw ndsher tsà*

If a man acts like mad or like a lunatic all of a sudden, it is believed that he is being selected to become a Llubbu or Sanyi (see page 150-151). It is believed that his acting like mad is due to his having been taken possession of by demons over which he has no control. In such a case a Tongba is called and the ceremony Päw ndshĕr tsà is performed. The mad man will be taken to the Sadon temple, Sadon being a protector of the Llubu, where the Tomba chant praying the spirit of Sadon to protect the incipient Llubu. A red cloth must then drop on him which becomes his turban and the mad man becomes a full fledged Llubu or Sanyi, the spirit of Sadon protecting him and thus enabling

* Power of god descend!

vol. 196.

187

him to call and meet and see demons unafraid. A lamp is lighted during the ceremony, a rooster, rice, sugar, ginger wine, tea and yellow paper are offered to Sadon. (White paper is only used for the dead).

Shĕr tsu la tsu

The believe is among the Nashi that wicked people harbor within them the demon Shĕr tsu la tsu that is the spirit or rather soul of a leopard and a tiger, when a person is thus said to harbor the demon Shĕr tsu la tsu it is able to give illness to people or cause the death of children. This may be compared with the witch business in the middle-ages. Shĕr tsu la tsu is said to be inordinately fond of gold and silver. A person who is being accused by others to harbor that demon will often call the Tombas to perform the ceremony of the above name, in order to get rid of that demon whom they are believed to carry about

Plants of the Coast of Coromandel

William Roxburgh, the 'Father of Indian Botany', studied medicine at Edinburgh where he was greatly influenced by the lectures of John Hope. He went to India as an East India Company surgeon, initially based in Madras, and set up experimental plantations for useful plants at Samulcottah in the Northern Circars of what is now Andhra Pradesh. During the 37 years he spent in India Roxburgh commissioned illustrations of some 2,500 Indian plants, the earliest ones while working as a surgeon and Company Naturalist on the Coromandel Coast, the later ones after he moved to take up the post of Superintendent of the Calcutta Botanic Garden. Roxburgh kept a version of each drawing for himself and sent a copy to the East India Company in London. Sir Joseph Banks, who advised the Company on botanical matters, realised the importance of these drawings and arranged for the publication of a selection of 300. The text was (anonymously) edited by his librarian Jonas Dryander, and the work issued in three elephant folio volumes between 1795 and 1820. The composition of the original drawings was rearranged during the process of engraving, and the watercolour washes (applied by hand to each print) are much less bold than the bodycolour of the original works. Shown here are *Roxburghia gloriosoides* (above, now *Stemona tuberosa*), a climbing plant with a generic name that commemorates Roxburgh, though now no longer recognised, and (inset) *Caesalpinia sappan*, the heartwood of which is the source of a red dye. **HN**

The Wallich Catalogue

Born in Copenhagen, Nathaniel Wallich (1785–1854) studied medicine, including botany. He joined the East India Company in 1809 and worked with William Roxburgh at the Company's Calcutta Botanic Garden. He engaged his own plant collectors and enlisted Company employees with botanical interests to systematically collect living and preserved plants. Duplicates of over 20,000 herbarium collections, representing some 9,150 species, were sent to more than 50 recipients in Europe and one in North America. Wallich used lithography to cheaply reproduce a listing to save him writing out specimen labels. The double-columned pages were printed on one side of thin, unwatermarked paper sheets of rather poor quality. Many of the recipients did not cut up the sheets, as Wallich intended, rather they bound them as a book at the end of the distribution. The Wallich Catalogue or, to give it its full title, *A numerical list of dried specimens of plants in the East India Company's Museum, collected under the superintendence of Dr. Wallich of the Company's Botanic Garden at Calcutta*, is of great importance as the basis for thousands of new species in Asian botany. With no more than 70 copies possibly in existence, probably many less, RBGE's three copies are the most held by any institution. **MW**

Himalayan Birch

Betula utilis

Robert Blinkworth's specimen of *Betula utilis*, the Himalayan birch, was collected in Kumaon in north-west India in the 1820s, while he was working as a plant collector for the East India Company. Nathaniel Wallich coined the name *Betula Bhojpattra* using the Nepalese vernacular 'Bhoj patra', a name he would have heard during his year in Kathmandu in 1820–1821. This high-altitude, stately tree is very useful – hence the scientific name – as the bark has medicinal applications and the branches are lopped for fodder. The wood is valued for construction and the papery sheets of peeling bark are used to line the roofs of houses in the Himalayan Mountains. The Far Western form (var. *jacquemontii*) has almost pure white bark and is popular in British gardens. **MW**

Blue Pea

Clitoria ternatea

At the request of botanists such as James Petiver, substantial numbers of Indian plant specimens were sent to London at the end of the 17th century, especially from Madras (Fort St George). These were used in taxonomic publications including those of the great naturalist John Ray. One of the major Madras collectors was Dr Edward Bulkeley (c.1651–1714), who sent this specimen to Charles Dubois in 1700. This specimen shows several interesting features, not least that some pigments, such as the blue of the flower, can last for three centuries, while others are fugitive. Also demonstrated is the extent and variety of nomenclature both in pre-Linnean European and indigenous Indian systems. Linnaeus based his generic name on a fancied resemblance of the flower to the female genitalia, but this was based on an earlier name in the work of Jacobus Breyn (1678), as seen from one of the names on this sheet. Dubois also cites a Sinhala name from Ceylon that is still used, though Bulkeley's transliteration of a Tamil name is not one of the 23 given in a recent list of south Indian plant names. **HN**

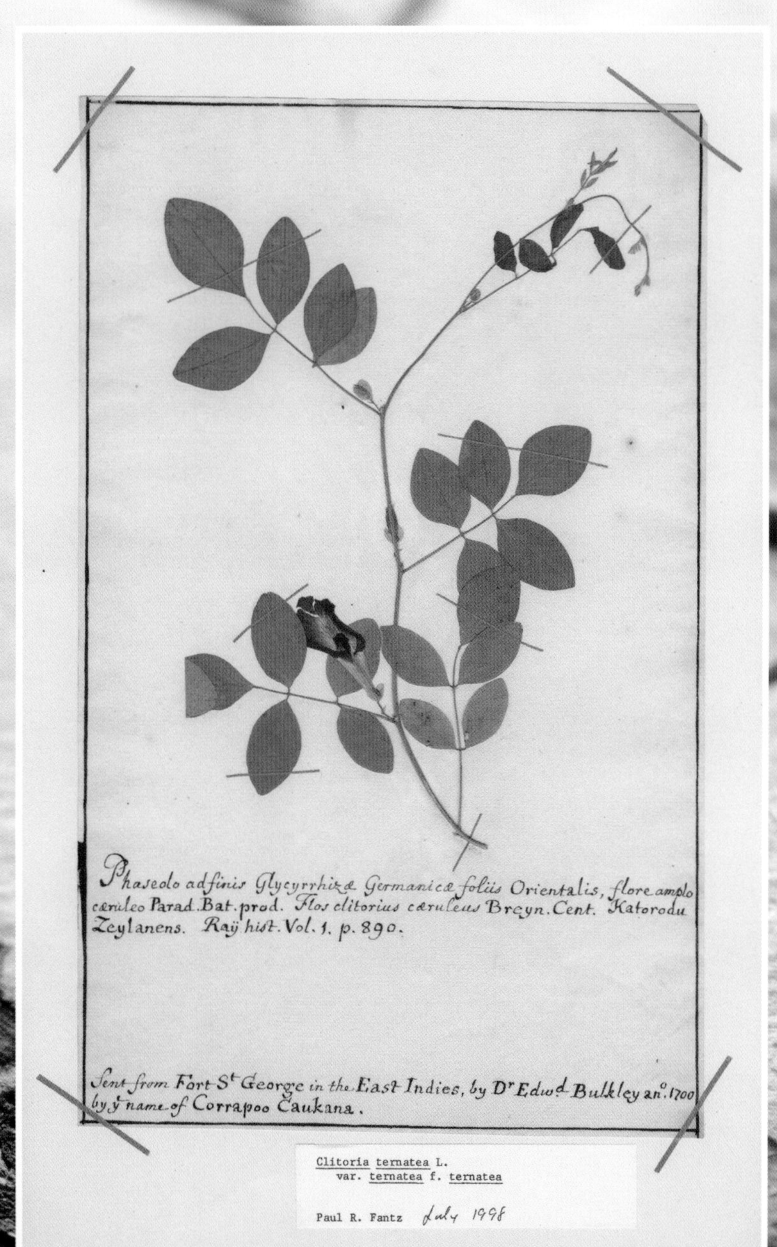

Phaseolo adfinis Glycyrrhizæ Germanicæ foliis Orientalis, flore amplo cæruleo Parad. Bat. prod. Flos clitorius cæruleus Breyn. Cent. Katorodu Zeylanens. Raij hist. Vol. 1. p. 890.
Sent from Fort St George in the East Indies, by Dr Edwd Bulkley ano. 1700 by ye name of Corrapoo Caukana.
Clitoria ternatea L.
var. ternatea f. ternatea
Paul R. Fantz July 1998

Wrightia arborea

While on the Coromandel Coast during his time in India, William Roxburgh started to describe plants and to have them illustrated by local artists (see p. 84); in 1789, he sent a herbarium of 565 species to the Royal Society in Edinburgh, including this specimen. He called the plant *Nerium tomentosum*, but cited an earlier illustration from Hendrik Van Rheede's *Hortus Malabaricus*, on which the name *Wrightia arborea* is based, so Roxburgh's name cannot be used. After his appointment to the Calcutta Botanic Garden, Roxburgh continued to have plants illustrated, the accompanying descriptions eventually published as *Flora Indica*. In this work he recorded that the yellow sap of this tree would dye cotton "a pretty good yellow" that was light-fast for at least two years. **HN**

Hexagonia wightii

This bracket fungus was described as *Polyporus wightii* in 1832 by the German botanist Johann Friedrich Klotzsch (1805–1860), who at the time was working in Glasgow as William Jackson Hooker's Herbarium Assistant. In 1838, it was transferred by the Swedish mycologist Elias Magnus Fries (1794–1878) to the genus *Hexagonia*. This specimen is probably part of the original material collected in southern India by the Scottish surgeon and botanist Robert Wight (1796–1872) of whom Klotzsch wrote: "We feel, therefore, gratefully committed to Dr Wight who, during his stay in Madras and Negapatam, sent people into the neighbourhoods at his own expense in genuine botanical spirit and taught them techniques of collecting and storing plants, when his duties as a doctor in the services of the East India Company did not permit him to do so himself. With a rich yield of both phanerogams and cryptogams, he returned to London last year to arrange his collection and to bring them to the light."

Polypore fungi display a wide range of pore shapes, from small and simple round pores (as in the red-belted bracket), to convoluted ridges in the oak mazegill. *Hexagonia* has regular yet more complex hexagonal pores, reminiscent of a honeycomb or the breeding chambers of common wasps. Inside the pores are structures producing millions of spores, through which the fungus disseminates itself. **SH**

Strobilanthes urticifolia

Even in the early 19th century a few women (usually well-born) managed to escape from the trammels of convention to become significant plant collectors, one such being Christian Brown Dalhousie (1786–1839), who signed herself ‘CBD’ on her specimens. Born Christian Brown at Coalstoun, East Lothian, she married George Ramsay, 9th Earl of Dalhousie in 1805. He was a soldier and she accompanied him on various postings, notably to Canada (where he was Lieutenant Governor of Nova Scotia) and to India, where he was Commander in Chief of the army. Lady Dalhousie collected this specimen in 1831 in the Western Himalaya, at Simla, which became the summer capital of the British Raj. This collection was described as the new species *Strobilanthes alatus* by the German botanist Christian Nees von Esenbeck, but this name had been used earlier. In 1837, she gave her collection of 1,200 Indian specimens to the Botanical Society of Edinburgh; it became one of RBGE’s earliest Himalayan collections. **HN**

Blue Vanda

Vanda coerulea

The expedition undertaken by Joseph Dalton Hooker to the Eastern Himalaya and north-east India from 1848 to 1851 was of pre-eminent scientific importance, and (from the *Rhododendron* species introduced) changed the face of British gardens. On part of the trip Hooker was accompanied by his friend from Glasgow University days, Thomas Thomson (1817–1878). This spectacular orchid had first been collected in the Khasia Hills (now in the Indian state of Meghalaya) in 1837 by William Griffith whose description of the flower colour caused great anticipation. Hooker and Thomson rediscovered it in October 1850 in the oak woods of Lernai "waving its panicles of azure in the wind", and collected seven porter-loads of living plants, which were then fetching between £3 and £10 per plant in London. Though not apparent from this specimen the colour in life is a vivid bluish-purple, with paler chequering, and it has been extensively used in hybridisation to make gaudy ornamentals. The orchid grows epiphytically, especially on oak trees, and occurs, though rarely, in north-east India, Yunnan (China), and northern Burma and Thailand. **HN**

Golden Flowered Dendrobium

Dendrobium chrysanthum

At the outset of the *Flora of Bhutan* project in 1975, it was decided that RBGE expeditions to Bhutan should focus on the southern half of the country, to explore the rich and still well-preserved forests neglected by earlier botanists in favour of the high mountains rich in cool-temperate and alpine plants suitable for British gardens. Late-night pressing of the day's collections was by the light of a Tilley lamp, a constant nocturnal companion due to the rarity of electricity in southern Bhutan. Amongst the collections were two epiphytic orchids, mailed back to RBGE – an acceptable practice at that time – in the hope that they would flower. This is one of them, which flowered in August 1982. Some of its flowers were pickled in spirit and although the colour is lost, the structure remains intact and can be invaluable for research. A flowering stem was also pressed to become a herbarium voucher specimen. **DL**

Reginald Farrer Watercolour

Reginald Farrer (1880–1920) was a polymath whose archive collections are now held at RBGE. A nurseryman, author of rock gardening and plant books, and a plant collector, Farrer undertook two expeditions, firstly to northern China between 1914 and 1915, and then to Burma (now Myanmar) in 1919. He became ill there and died in 1920. In addition to his botanical work he tried his hand as a playwright, novelist and politician, and was a correspondent during the First World War. He also painted, and many of his watercolours are now in RBGE's collections. During a time when the first glimpses of new plants from the Orient were arriving in Britain in the form of dried plants, seeds or sepia photographs, to have views of them clinging to the cliffs in their natural habitat in full colour must have been quite novel. **LP**

The George Forrest Collection

Born in Falkirk, George Forrest (1873–1932) is one of Scotland's most celebrated plant collectors. On his return to Scotland from travels in Australia and South Africa in 1902, he obtained a post in the RBGE Herbarium where his acquaintance with Isaac Bayley Balfour led to him being recommended for a collecting trip to the mountains of Yunnan, south-west China, in 1904. This was the first of seven expeditions Forrest undertook, collecting more than 30,000 plant specimens and visiting regions where political and religious tensions made collecting particularly perilous. In one letter Forrest talked of spending almost three weeks evading locals with murderous intent. The Forrest collection, held in the RBGE Archive, provides an insight into both life in China in the early 20th century and the often challenging nature of plant collecting. **LP**

Plum Blossom

Prunus mume

Travelling through the remote areas of Yunnan, Sichuan, Tibet and Upper Burma, George Forrest discovered over 1,200 new species whilst collecting more than 30,000 specimens. Philipp Franz von Siebold first described this species in 1830, noting that it had both white and pink flowered forms. This creamy-white flowered specimen was collected in Yunnan by Forrest in 1913, a year before the start of the First World War. Intriguingly, *Prunus mume* is now celebrated as the national flower of Taiwan, first designated on 21 July 1964, only a few weeks after the opening of the new Herbarium and Library building in Edinburgh. **EH**

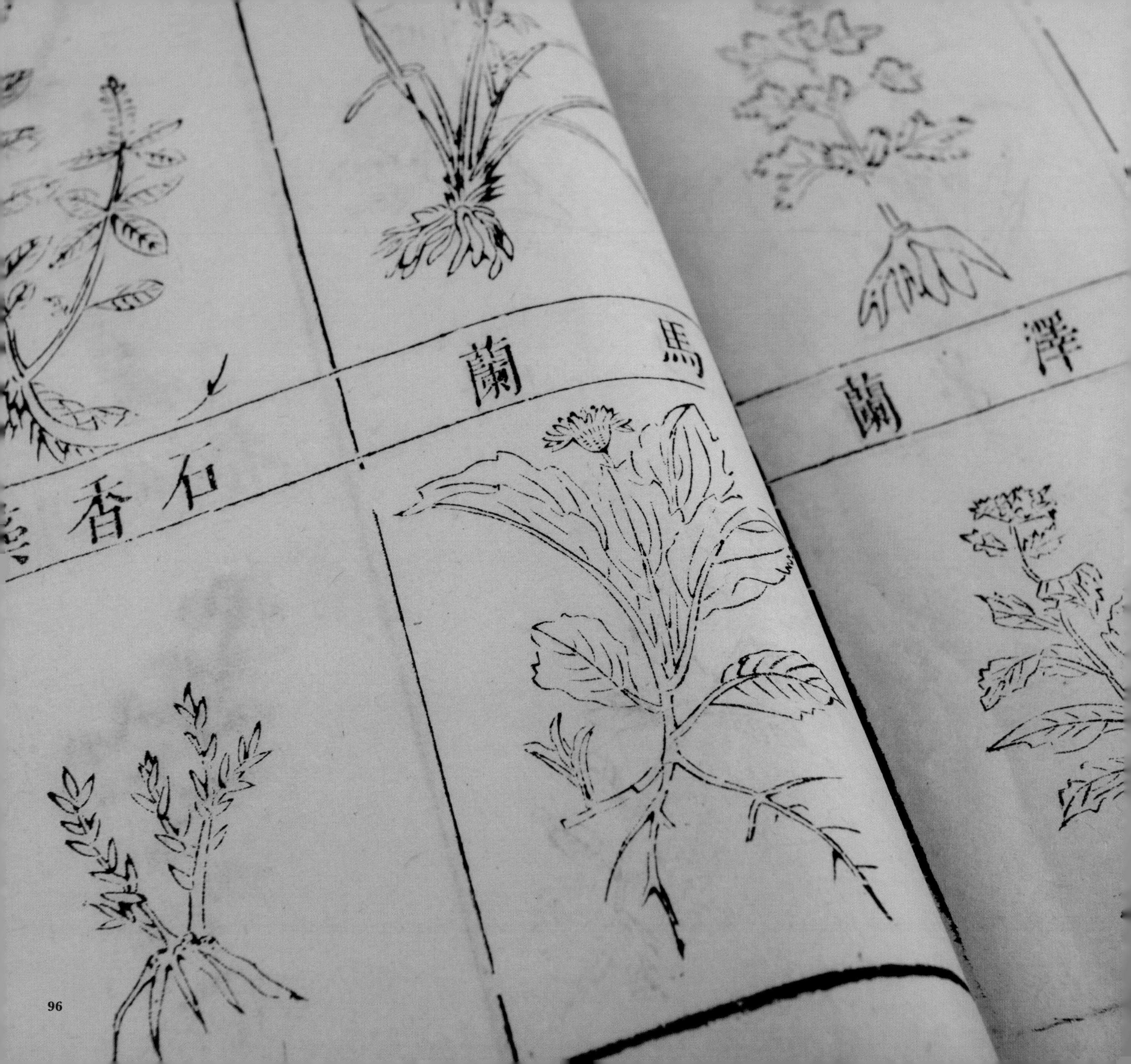
馬蘭
石香
澤蘭

Pen Ts'ao Kang Mu

This, the most important of early Chinese works on materia medica and natural history, was compiled by the civil servant and district magistrate Li Shih-Chen between 1552 and 1578 and first published, posthumously, in 1596. The earliest surviving edition is from 1658, and this copy, which belonged to John Hope and had been given to him by his pupil William Hamilton (1758–1790), Professor of Botany at the University of Glasgow, may date from this period. Like the Latin *Herbarius* (p. 16), the woodcuts were copied from earlier works and are correspondingly, as noted by Emile Bretschneider, a 19th-century scholar of Chinese botanical history, "so rude that it is very seldom that any conclusion can be drawn from them". Hope's copy represents only one section of the lengthy work and includes illustrations of minerals as well as plants. This is by far the earliest Chinese printed work in the Library, and prefigures the great 20th-century interest of RBGE that culminated in our contribution to the *Flora of China* (25 volumes of text and 22 of illustrations) published between 1994 and 2013. **HN**

From my esteemed friend Professor Willm Hamilton of Glasgow

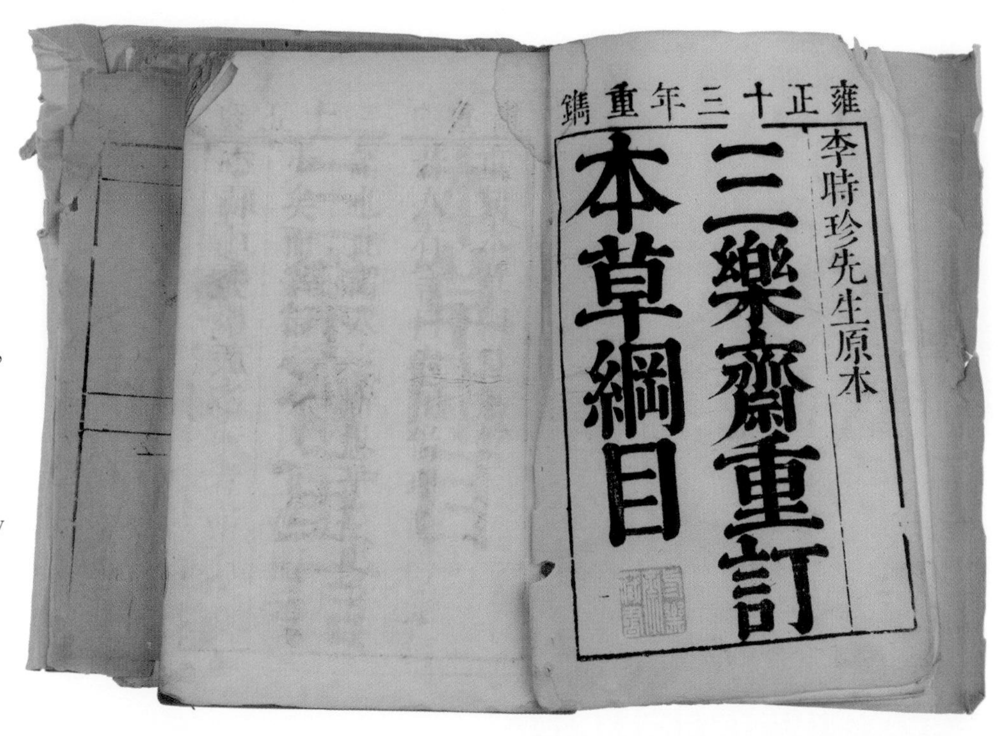

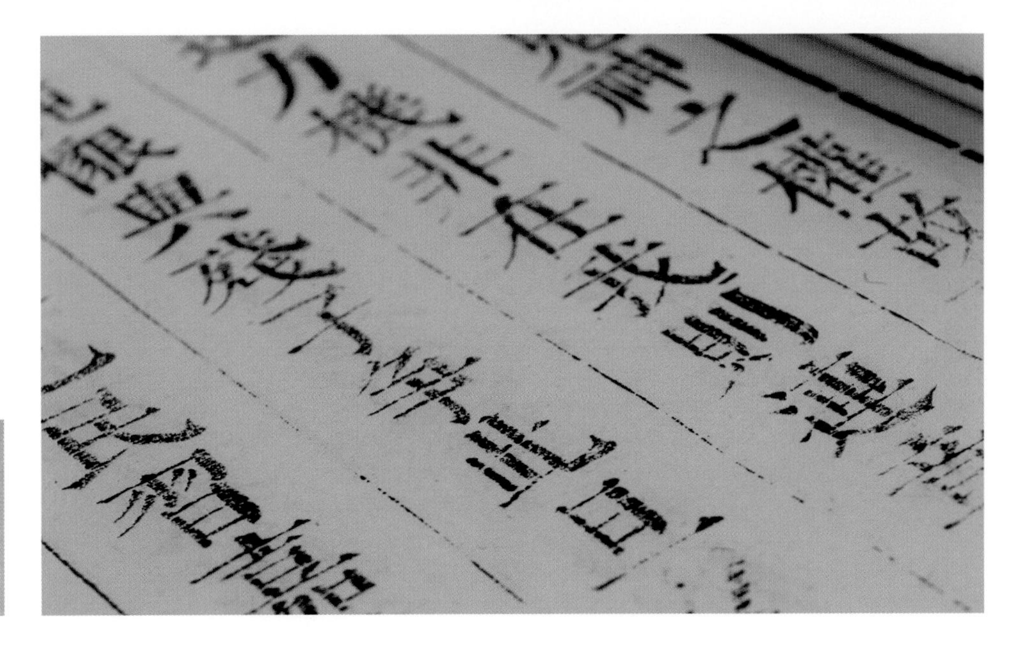

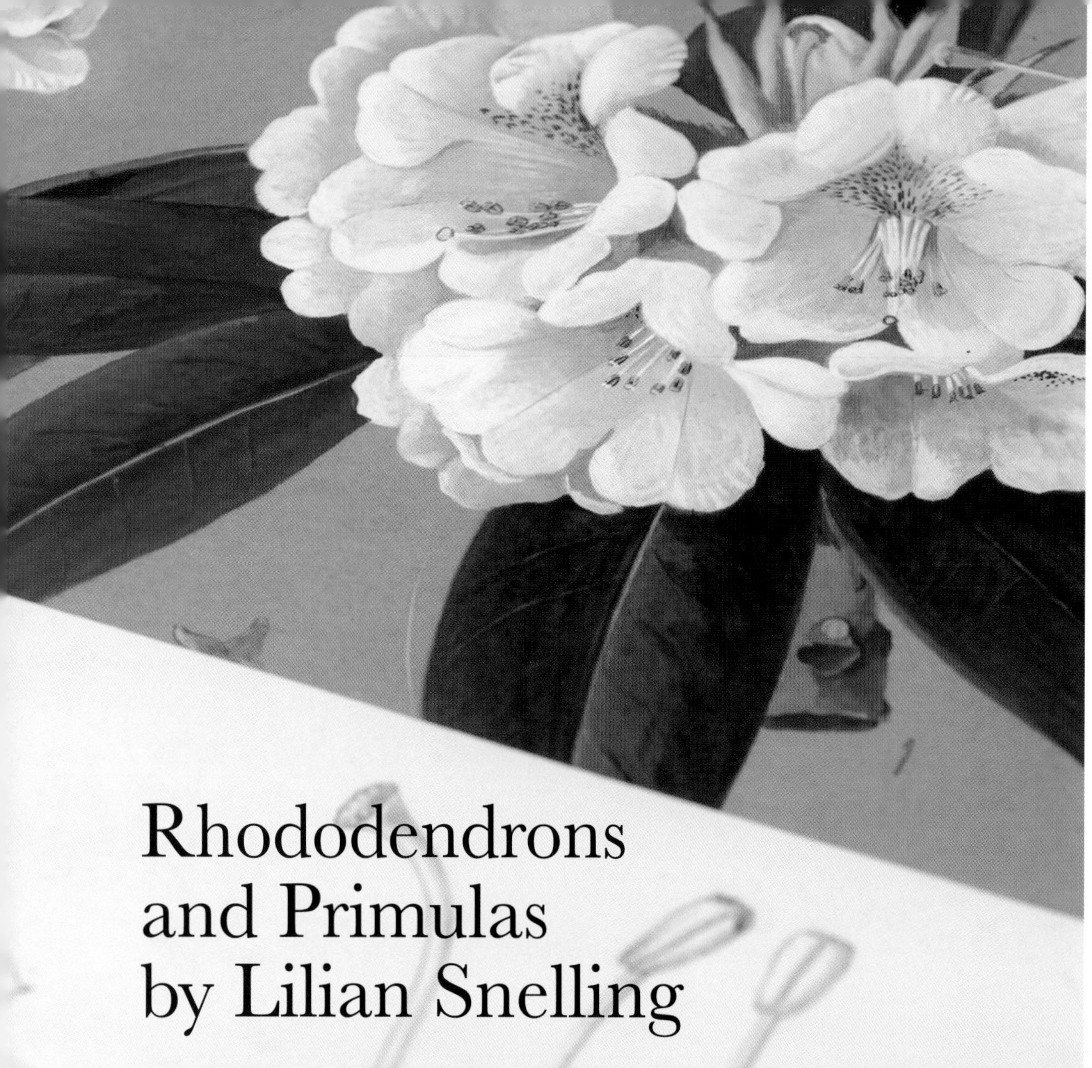

Rhododendrons and Primulas by Lilian Snelling

One of the most popular and highly regarded of the botanical artists represented in the RBGE art collection is Lilian Snelling (1879–1972). Born at Spring Hall, St Mary Cray in Kent, Snelling first came to prominence in 1915 when the plant collector Henry John Elwes commissioned her to paint the flowers at his home in Gloucestershire. In 1916, Snelling moved to RBGE and developed the botanical accuracy of her paintings under the demanding eye of Isaac Bayley Balfour. Much of this work involved capturing microscopic differences between species of plants, with the drawings used as working documents by RBGE scientists, some becoming heavily annotated. In 1921, Snelling left Edinburgh to take up a post with the Royal Horticultural Society as both principal artist and lithographer on the journal *Curtis's Botanical Magazine*, a post she held for 30 years until her retirement in 1952. Snelling was awarded the Victoria Medal of Honour, the highest horticultural award in Britain, in 1955. **LP**

X2
X10
x2
Thrum-eyed flowers.
X8
X2
X2
on upper surface
and edge of leaf.
Pin-eyed
on underside
of leaf
x2
X10
x5
Pin-eyed plant
L.S.

Lampshade Poppy

Meconopsis integrifolia

Although the genus *Meconopsis* is best known for its blue-flowered members, there are also species with pink, white and yellow flowers. Among the last, with flowers up to 25 centimetres in diameter, is what Reginald Farrer called the "lampshade poppy", which, before flattening in the flower press, are "sulphur-pale orbs". First discovered by the Russian naturalist and explorer Nikolai Przewalsky in Gansu in 1872, it is widespread in western China, occurring from Gansu and Qinghai through eastern Tibet to western Sichuan and Yunnan, at altitudes of over 2,800 metres. This specimen was collected in Sichuan by a Chinese collector working for the Hon. Henry Duncan McLaren, later the 2nd Lord Aberconway, an industrialist, politician, art collector and horticulturist, who created a notable garden at his family home Bodnant Hall in north Wales. McLaren had contributed to George Forrest's last two expeditions, and realising the value of the team of Chinese collectors trained by Forrest, continued to pay them after Forrest's death, which occurred in 1932. **HN**

FLORA OF Szechuan
NAME Meconopsis integrifolia
LOCALITY Tatsienlu
HABITAT Summit of a mountain ALT.
REMARKS Herbaceous flower, colour yellow
Blossom in June
COLLECTOR McLaren
REF. No. AC 8
DATE
Meconopsis integrifolia (Maxim.) Franch.
C. C. Yuan
July 2002
ZHONGSHAN (SUN YATSEN) UNIVERSITY

Primula sonchifolia

One of the major 20th-century botanical interests of RBGE was the genus *Primula*, which reaches its greatest diversity in the Sino-Himalaya. They are small herbs with disproportionately large and colourful flowers, many of which thrive well in the cool, damp conditions offered by Scottish gardens, though short-lived and requiring meticulous cultivation. The taxonomy of the large genus (some 430 species) is intricate and was studied at RBGE by a unique combination of skilled collectors/horticulturists, Roland Cooper and George Forrest, working with a succession of three Regius Keepers, Isaac Bayley Balfour, William Wright Smith and Harold Roy Fletcher (1907–1978). This culminated in a multi-part monograph by the last two authors published between 1941 and 1949. The specimen shown here reminds us that most Western collectors operating in foreign parts, with limited knowledge of local languages and topography, relied heavily on local expertise, though this was seldom acknowledged. The detailed habitat and locality information (translated on the left) on this label was written by an anonymous Chinese collector trained by Forrest and later used by McLaren. **HN**

Primula sonchifolia Franch.
Yunnan 1934. (362)

McLaren L collection.
29A
herbaceous flower
storied
blue flower
It is on the top of Lien Wang Mt. of Chao Chou
blossom in July

T' 29A
草花
樓台藍花
趙州界連王山頂
柒月開花

Eve Reid Bennett's Paintings

Eve Reid Bennett first came to RBGE in 1989 having studied design at Edinburgh College of Art and painting and printmaking in Vancouver. After her return to Scotland, her own garden inspired her to take up botanical painting and she began a long association with RBGE, during which she taught the first classes in botanical illustration. After six years of teaching she decided to concentrate on her own painting, working closely with the botanist and plant collector George Argent to illustrate the results of his 35 years of *Rhododendron* collecting in South East Asia. Shown here are three species of *Vireya* rhododendrons drawn from specimens collected by George Argent. Also, on the far right, is a drawing of *Tacca chantrieri* – Bennett was fascinated by the shapes of the complex inflorescences of *Tacca* species, also known as bat lilies or cat's whiskers. **LM**

probable ISOTYPE
of DACRYDIUM TAXIFOLIUM
Banks & Sol. ex D. Don
Ex Herbario Musei Britannici.
Podocarpus spicatus Br.
No. NEW ZEALAND.
Banks & Solander; Cook's first Voyage. 1768-71.
ROYAL BOTANIC GARDEN
EDINBURGH
E00127683
Prumnopitys taxifolia (Banks & Sol. ex D. Don) de Laub. (= Pod. spicatus Br)
Determinavit R. R. Mill
3/8/97.

Black Pine

Prumnopitys taxifolia

This specimen was collected in New Zealand between 1768 and 1771 during the voyage of the *Endeavour*, the first of three voyages to the South Pacific under Captain James Cook. This collection was made by the botanists Joseph Banks and Daniel Solander, and like many of the 350 plant collections they made from New Zealand, this conifer was new to science. Although *Prumnopitys taxifolia* occurs on both the North and South Islands, there are now very few examples of old-growth forests left due to widespread logging. RBGE holds over 15,000 conifer herbarium specimens collected from around the world; this specimen is one of 165 which represents original (type) material. Such specimens provide important baseline information for taxonomic research. Today the main focus for such research concerns the monographing of the Podocarpaceae, the family to which *Prumnopitys taxifolia* belongs. **MG**

Cardwell Lily

Proiphys amboinensis

Perhaps the greatest misfortune visited upon Malesian botany is the untimely death of the Aberdonian botanist William Jack (1795–1822). Jack was a remarkable individual, a most precocious scholar and outstanding natural historian. He left for India in 1813 on his 18th birthday, landing on Sumatra in 1819 to work under Sir Thomas Stamford Raffles (1781–1826) on the natural, geological and cultural history of the island. Jack's extensive herbarium from the region was destroyed during a tragic fire aboard the *Fame* in 1824, after his death from malaria. This specimen is one of the few Jack collections which made it back to Europe, and it turns out to have been sent by him as a prank. The specimens were requested by Lady Hastings, who was not highly thought of by the collector. Jack stated that he had made "a parcel of second rate ones, with plenty of good paper, which is of more consequence … I trust her indolence never to look at them; indeed if she did I don't suppose she would know a mangosteen from an apple …" **MH**

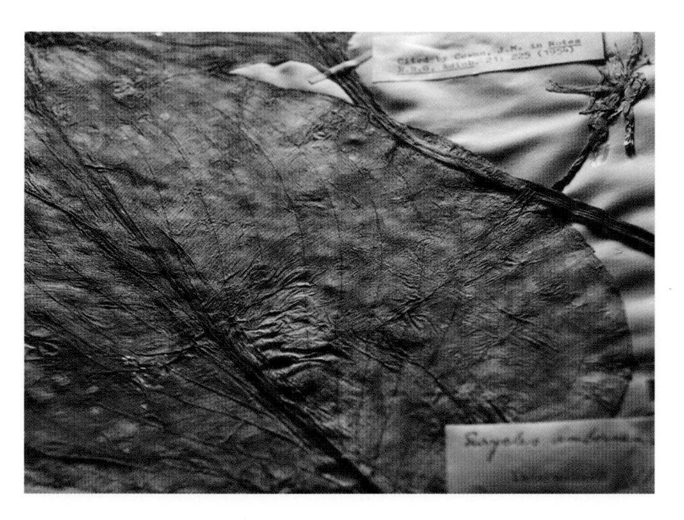

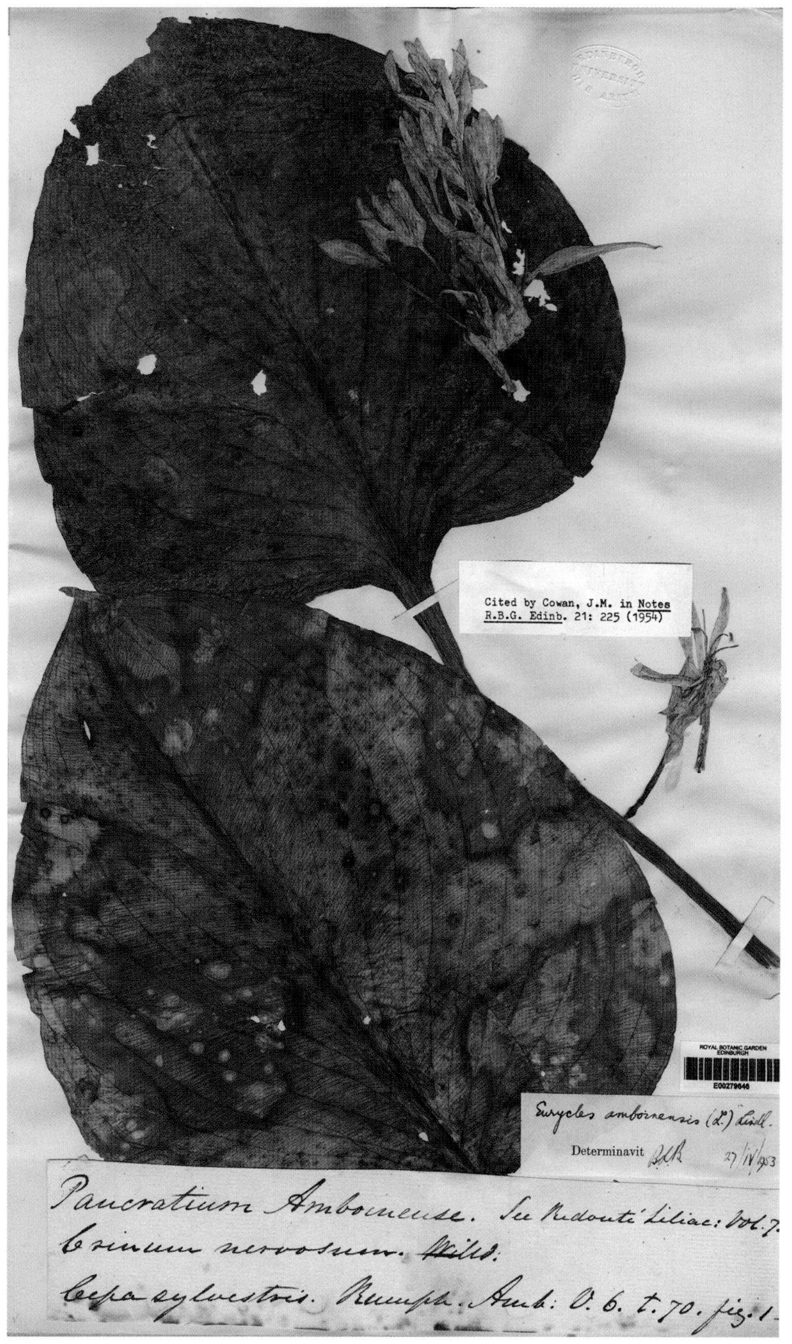

Bird’s Nest Banksia

Banksia baxteri

Robert Brown’s exploration of Australia at the beginning of the 1800s marked one of the most exciting and productive expeditions for the discovery of new species. The eminent Scottish botanist collected over 3,000 specimens, of which over half were considered to be new to science. Born in Montrose in 1773, Brown studied medicine at the University of Edinburgh, but became more fascinated by botany and spent much of his time botanising in the Scottish Highlands. Following the expedition to Australia he continued his career in London, becoming the first Keeper of the Botanical Department at the Natural History department of the British Museum (now the Natural History Museum, London). During his studies of pollen grains he also discovered the motion of particles now known as Brownian motion. When he died Brown bequeathed his herbarium specimens to his assistant, J.J. Bennett, who catalogued them. The specimens here were received in 1877–1878 on the directions of Bennett. **EH**

R. Brown, Iter Australiense, 1802–5.
[Presented by direction of J. J. Bennett, 1876]
No
Banksia Baxteri, Br?

Presented by
Dr. W. Lauder

Economic Botany

The relationship between humans and plants is possibly more strained now than at any other time in its long history. While we depend on plant resources in many aspects of our lives, knowledge of the plants which lie behind these resources is held by a shrinking percentage of the population. In addition, the transplanting of species as crops around the world has often obscured the origin of some of our most popular commodities. This selection shows just some of the diverse uses of plants and some of the stories involved in their discovery and relocation.

A Curious Herbal

Aberdeen-born Elizabeth Blackwell (1707–1758) was both the artist and the engraver for *A Curious Herbal*. Aware of the need for an English-language herbal for training apothecaries, she used her skills and connections, and her husband's medical and linguistic knowledge, to produce a book to meet that need. She commenced the work in 1737 and, by 1739, 125 parts, describing and illustrating four plants each, had been produced. Taking lodgings near the Chelsea Physic Garden gave Blackwell access to an exceptional plant collection; access to the collections of patrons, including Sir Hans Sloane, enabled her to include exotic species such as tobacco, sassafras, tamarind and date palm in the herbal, alongside common British plants such as foxglove and dandelion. Volume 1 of the RBGE copy is inscribed by John Hope to Sir James Naesmyth (1704–1779), 2nd Baronet of Dawyck and Posso, botanist and silviculturist, "Doctor Hope prays Sir James Naesmyth will allow this Ladies performance a place among his Botanical works", and may have been returned to Hope when Naesmyth died. **GH**

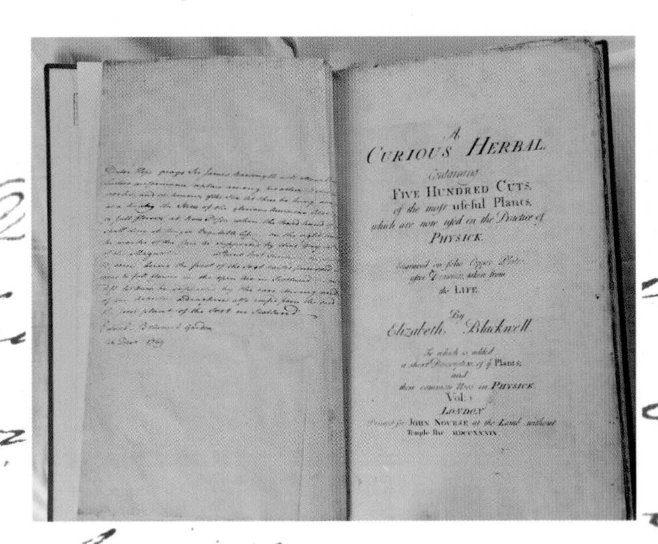

A
CURIOUS HERBAL,
Containing
FIVE HUNDRED CUTS,
of the most useful Plants,
which are now used in the Practice of
PHYSICK.
Engraved on folio Copper Plates,
after Drawings taken from
the LIFE.
By
Elizabeth Blackwell.
To which is added
a short Description of ye Plants;
and
their common Uses in PHYSICK.
Vol. I.
LONDON:
Printed for JOHN NOURSE at the Lamb without
Temple Bar. MDCCXXXIX.

Breadfruit

Artocarpus altilis

The first mission of Captain William Bligh to transport breadfruit plants from their native South Sea Islands to the Caribbean, where they were to provide much-needed food for the slaves, was famously unsuccessful, leading to the mutiny on the ship, the *Bounty*. However, the second mission in 1791 succeeded and the species was successfully planted in St Helena, St Vincent and Jamaica, although the slaves would not eat the fruit. The specimen shown here was taken from one of Bligh's original plants. The locality is recorded as Mauritius which raises the intriguing question of when the plants were relocated to this island in the Indian Ocean. **EG**

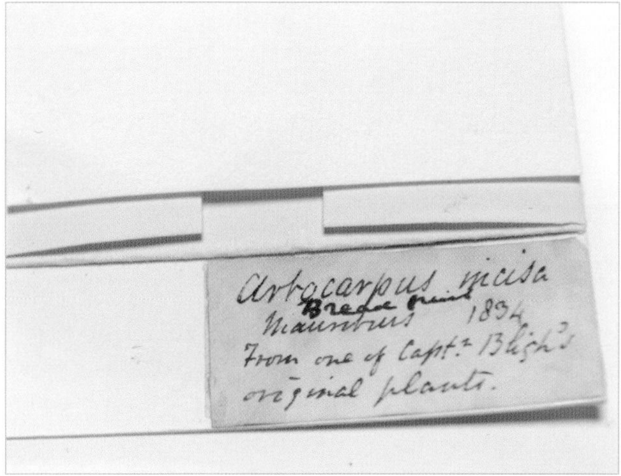

Opium Poppy

Papaver somniferum

Though still one of the most important sources of painkilling drugs, the products of the opium poppy have always been subject to misuse. None more so than in the shameful activities of the East India Company in the 19th century, when it was profitably and cynically used as an item of trade for Chinese tea. One of the great areas of poppy growing was northern India (Bengal and Bihar), and this specimen was grown at RBGE from seed from Ghazipur, where an opium factory had been established by the Company in 1820. This factory is still in operation, producing opiates for the pharmaceutical industry, though under high security breached only by a troop of junkie monkeys; it is the largest such factory in the world, covering 17 hectares. This specimen comes from James McNab's demonstration herbarium, and would have been used in the lectures of his boss John Hutton Balfour. Interestingly, Balfour's first published paper was on a brain tumour that killed the son of perhaps the most famous opium eater, Thomas de Quincey. **HN**

Quinine Tree

Cinchona pubescens

Quinine, extracted from various *Cinchona* species native to Peru and surrounding countries, was for centuries the only effective treatment against malaria. The 19th-century battle to break the South American monopoly, so that it could be grown in Asia (in British India and Dutch Java), is one of derring-do and phyto-espionage. Ultimately the Dutch were the more successful, with the high-yielding *C. ledgeriana*, but the 'red bark' *C. succirubra* (now *C. pubescens*) was extensively grown in India. It had been collected for Kew by Richard Spruce and Robert Cross and taken to India by the latter. Thomas Anderson, an Edinburgh graduate who became Superintendent of the Calcutta Botanic Garden, set up plantations at Mungpoo near Darjeeling, where this specimen was collected by Andrew Jaffrey. Jaffrey had been a gardener at the Caledonian Horticultural Society's experimental garden under James McNab (from whose herbarium this specimen comes). In 1853, Jaffrey was sent to India and ended up working in the Darjeeling quinine plantations. **HN**

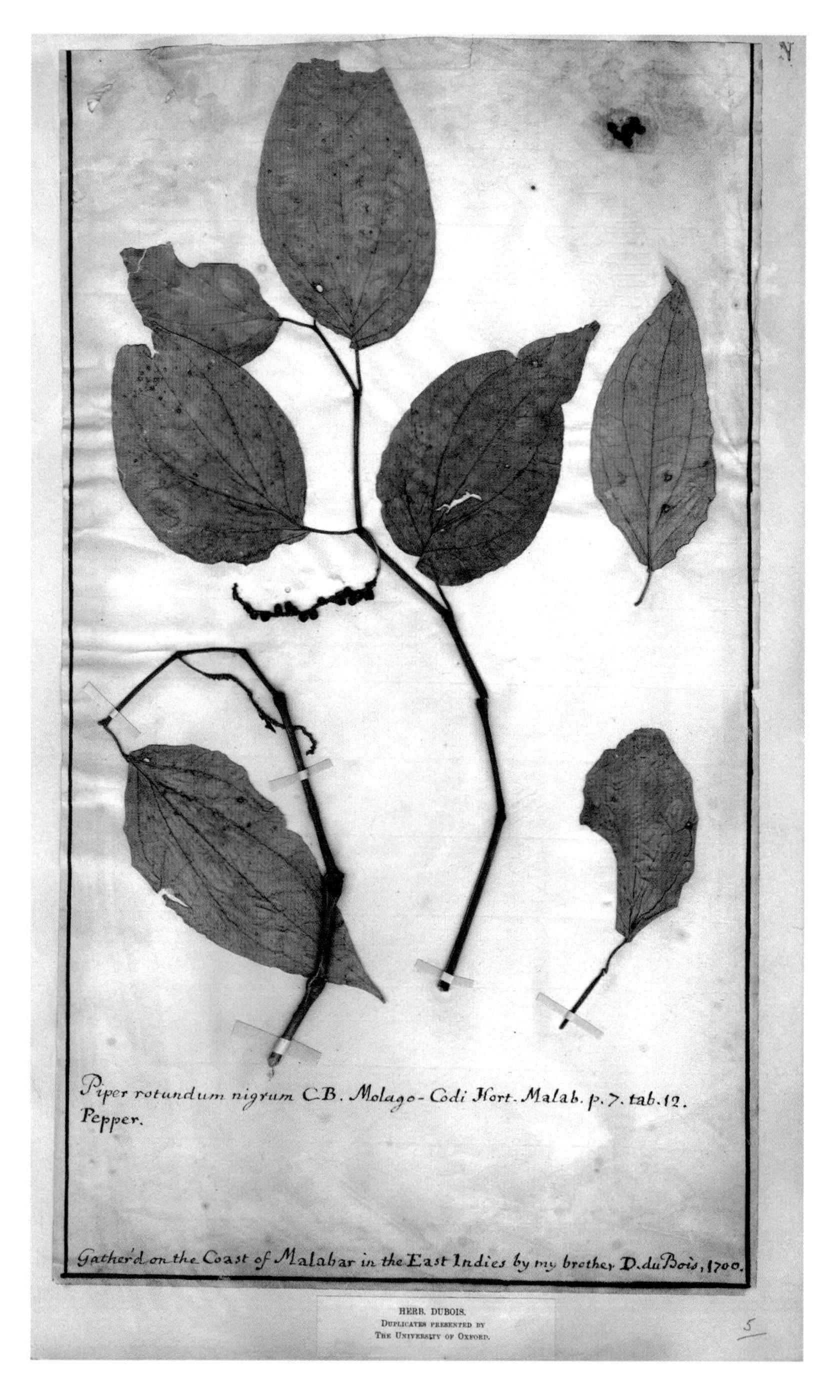

Piper rotundum nigrum C.B. Molago-Codi Hort. Malab. p. 7. tab. 12.
Pepper.
Gather'd on the Coast of Malabar in the East Indies by my brother D. du Bois, 1700.
HERB. DUBOIS.
Duplicates presented by
The University of Oxford.
5

Black Pepper

Piper nigrum

Since Roman times black pepper, native to Malabar on the west coast of India, has been an item of international commerce; it was a major incentive to colonisation by the Portuguese from the 15th century onwards. This specimen was collected in 1700, in Malabar (now the state of Kerala), by Daniel duBois (d. 1702), a merchant and at one point Attorney General of Madras, a British settlement on India's eastern Coromandel Coast. He sent the specimen to his half-brother Charles, Treasurer of the East India Company, in London. The annotation refers to an illustration in the seventh volume of *Hortus Malabaricus*. This work is one of the classics of tropical botany, compiled by the Dutch Commander of Malabar, Hendrik van Rheede, who co-ordinated a team of Indian plant collectors, avyurvedic doctors, translators and Dutch military draughtsmen. The work was published in 12 volumes in Amsterdam between 1678 and 1693; in it plant names are given in three Indian languages including Malayalam, as in the name 'Molago-Codi'. **HN**

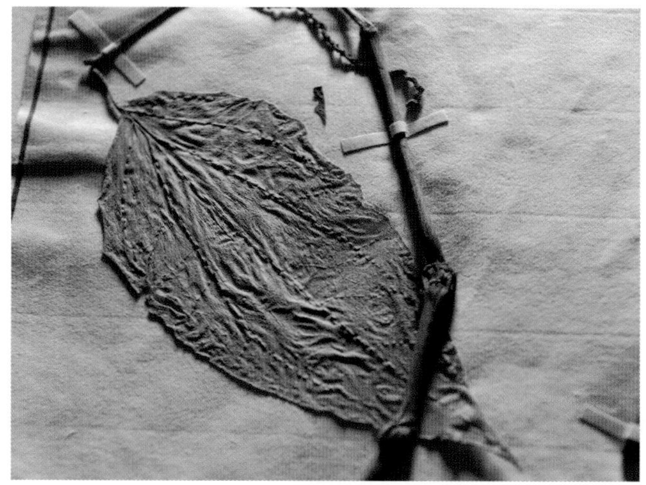

Coffee Rust

Hemileia vastatrix

As well as being harmful to plants and taking a toll on food harvests, plant diseases have changed human history. In the 1840s, the late blight of potatoes caused death and misery, leading to a decline in the population of Ireland by some two and a half million people. A destructive pathogen had decimated a susceptible crop in less than three seasons. In similar circumstances, but on a less dramatic scale, the rust fungus of the coffee plant, *Hemileia vastatrix*, invaded the coffee plantations of Ceylon in the 1860s, bringing the production to an end by the 1890s. In the wake of the disease the cultivation of other crops was attempted, notably cocoa and cinchona, but this proved unsuccessful. By 1880, the first commercial tea plantations were established and production soon increased, overtaking coffee by the end of the decade. It is said that the British taste for tea in preference to coffee stems from this change of cultivation in Ceylon. Since its run across Ceylon, coffee rust has spread to coffee plantations in India, Africa and more recently South America. The Herbarium holds a specimen of *Hemileia vastatrix* from Ceylon from 1880. As is the case with all rust spores, they are beautiful when seen at high magnification by electron microscopy. **SH**

Hemileia vastatrix on C. Arabica.

Hemileia (B&B):-
Sori somewhat circinating, hypophyllous, naked: flocci distinct, inarticulate, flexuous; spores reniform, at length granulate, verrucose on one side, obliquely attached at base by papilla.

H. vastatrix (B&B):-
White, orbicular patches below – burnt above. Spore smooth on side next flocci

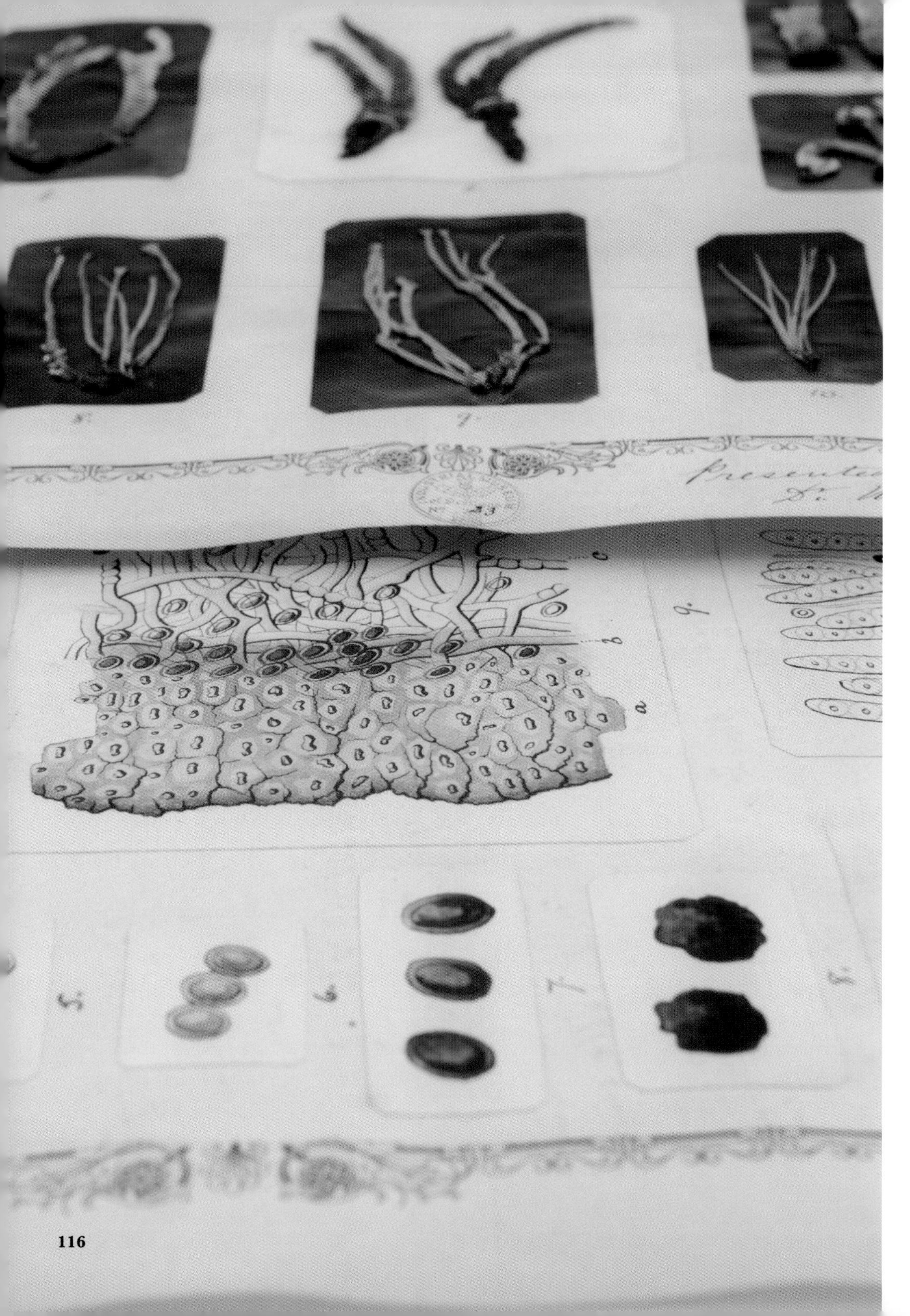

Lichens for Dye Collection

William Lauder Lindsay (1829–1880) was primarily a physician, employed at the Murray Royal Institution for the Insane in Perth, where he was recognised as an innovator and careful student of psychiatry. However, John Hutton Balfour, Regius Keeper of RBGE, encouraged Lindsay's interest in the study of lichens, which ultimately led to the latter winning the first Neill gold medal from the Royal Society of Edinburgh for a popular work on the subject. Lindsay was not content to merely describe lichens to interested naturalists; his very practical aim was to promote the potential of the Scottish economy in the Highlands, where living conditions were of a very poor standard. He performed hundreds of experiments on common and abundant lichens in order to find economical replacements for the expensive and increasingly scarce foreign dye materials, which, he noted, could fetch up to £1,000 per ton in the London market. He was a prolific writer on the subject and brought together the nucleus of a collection, displaying the lichens and their potential as dyes for exhibition. Here is a sample of the lichens and fabric swatches with heavily annotated sheets describing his methods and the lichens involved in producing the dyes. **RY**

Chinese Indigo

Persicaria tinctoria

Natural forms of indigo (best known for the dyeing of denim) are extracted not only from the genus *Indigofera* in the family Leguminosae, but from plants belonging to several other families, including woad in the cabbage family and this plant in the dock/knotweed family. Chinese indigo is native to China and Vietnam, but also widely cultivated there and in Korea and Japan.

This specimen was gathered in 1913 by the redoubtable American collector Mary Strong Clemens (1873–1968) in the province of Shanxi, northern China. Joseph Clemens, her husband, was a US army chaplain and together they made extensive collections, notably in the Philippines, China (the provinces of Shanxi and Hebei, then called Chihli), Borneo (especially Mount Kinabalu) and New Guinea. Joseph died of food poisoning in New Guinea in 1936, where Mary continued to work until being forced to move to Australia in 1941. In Australia she continued botanising into her 90s, working in the Queensland Herbarium, Brisbane, maintaining her spirits (to the annoyance of neighbours) by the vigorous singing of hymns. **HN**

Fairy Primula

Primula malacoides

This commonly grown conservatory plant counts as one of George Forrest's most successful introductions, though it had first been collected in 1884 by Jean-Marie Delavay, a French missionary to China. It is believed that all the plants in cultivation are derived from the seed collected with this herbarium specimen, made at Dali, on Forrest's first expedition in 1906. This material first flowered in the West in 1908, after which a huge and rapid diversification took place – in terms of colour, scent and vigour. The plant flowers in four months, but in susceptible humans can have unfortunate allergenic effects. In the wild, where it occurred mainly in Yunnan (straying into northern Burma and western Kweichow), the plant has not been seen recently. Unusually for a primula it is winter flowering (October to March) and its main habitat was as a field weed. Forrest noted that it was spread with the transport of bundles of crops such as beans, so it may have succumbed to improved cultivation techniques. **HN**

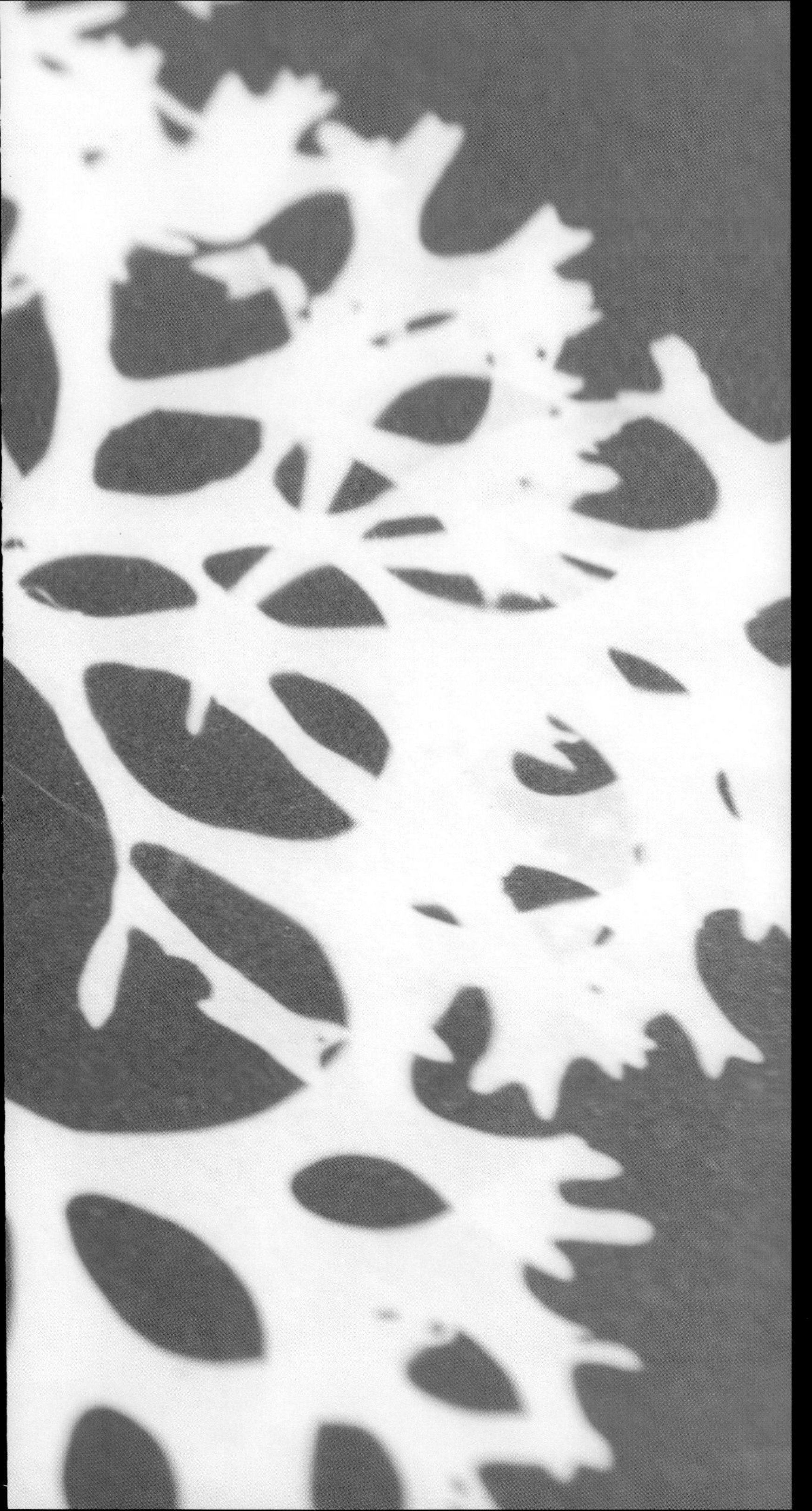

Art and Nature

These items show some of the ways that people have found to represent nature using printing and photography. As some of the earliest examples, they are of particular interest and importance. Herbarium specimens are often decorative in themselves, with beautiful forms, textures and occasionally colour; these characteristics can often be enhanced through careful and imaginative mounting. Of course it is always possible to enhance the natural beauty of specimens with ornate decorations, which in some cases can suggest that the plant itself is a minor element in the display.

D. IO. HIERON.
PATHOL. ET PRAX. IN ACAD.
ORDIN. FACVLT. MED. SENIOR. ET
ACAD. CAESAR. NAT. CVRIOSOR.
ET BIBLIOTHECARII
BOTANICA IN
SEV
IN
PLANTARVM
QVAM
PECVLIARI
ATRAMENTO

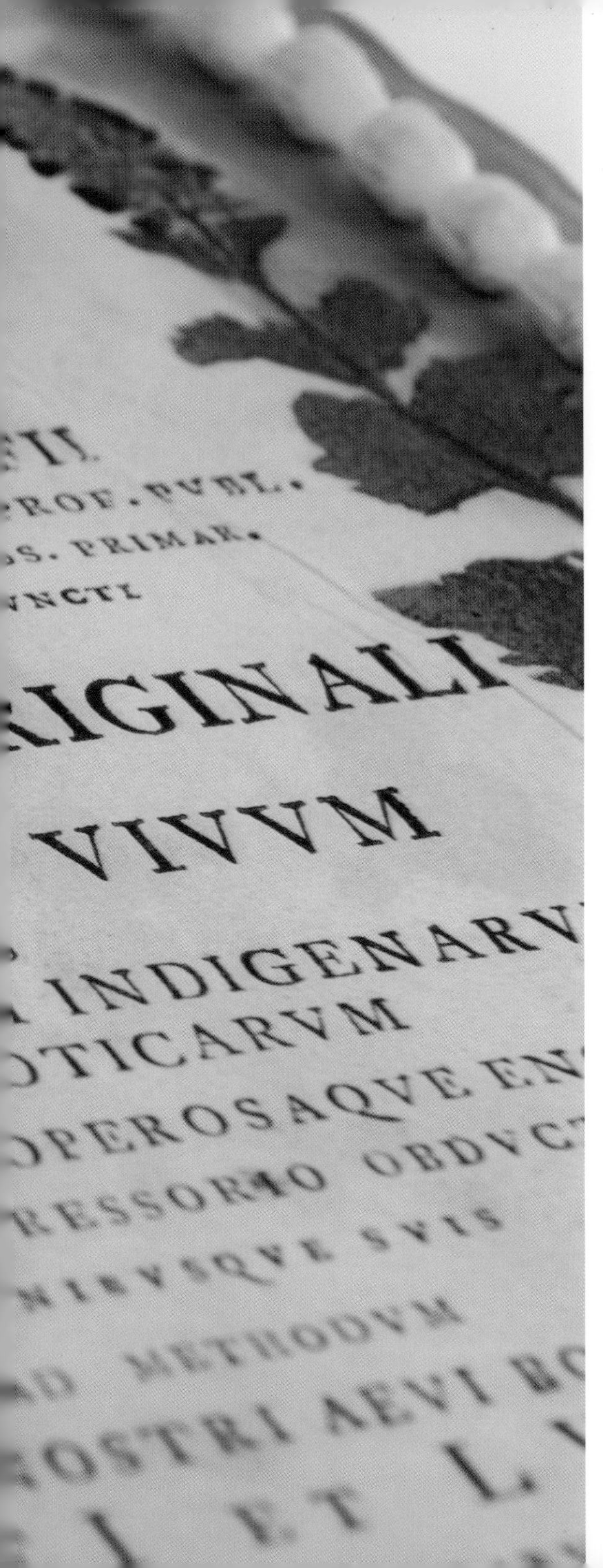

Botanica in Originali seu Herbarium vivum

The earliest known European nature prints (illustrations created by fusing an impression taken from actual plant material on paper) date from 1228, but the first known description of a process for nature printing dates from the 16th century. By the 18th century, nature printing using a printing press was developed in Germany. The instigators of this development were a university printer and a young medical faculty member in the city of Erfurt, Johann Michael Funcke and Johann Hieronymus Kniphof. They published two works using nature prints in 1733 and 1748–1749. From one prepared plant specimen 12 nature prints could be taken and these were then hand-coloured in the print shop.

Kniphof remained in Erfurt, but moved printing operations to Halle in the 1750s, where work began on a large-scale edition of *Botanica in Originali seu Herbarium vivum* (an earlier version of this work had been published in Erfurt). This eventually grew to 12 parts, each depicting 100 plants. When Kniphof died in 1763, editorial responsibility passed to Friedrich Wilhelm von Leysser and the work was completed in 1764. This is one of the earliest books to cite botanical names from Linnaeus's *Species Plantarum* (1753). The RBGE set contains all 12 parts, bound in six volumes. Although it has no ownership marks it most likely came from the library of John Hope. In 1899, when Hope's library was received by RBGE in a bequest from his grandson, an entry in the list of books that was drawn up reads: "Kniphof's *Herbarium Vivum*, 6 volumes." **GH**

Portfolio of Bradbury Nature Prints

At the Great Exhibition of 1851, Henry Bradbury, eldest son of the senior partner in the London printing house of Bradbury and Evans, became interested in the displays of nature printed works from the Austrian Staatsdruckerei. So much so that in 1852, aged 21, he went on a fact-finding mission to Vienna. In his desire to be the authority on nature printing in Britain, Bradbury used dubious business practice; his Austrian hosts were not pleased when he took out a patent on the process and an acrimonious pamphlet war ensued. Tragically, Bradbury committed suicide shortly before his 30th birthday. This history is a sad backdrop to the beautiful images produced by the Bradbury and Evans nature printing process. These plates are a few samples from Bradbury's first nature printed work, *A Few Leaves Represented by Nature Printing* (1854). This is a scarce work, and none of the copies known to exist today contain exactly the same combination of plates. **GH**

Nature Printed Ferns

Henry Craven Baildon was an active figure in Edinburgh's scientific and medical circles, a champion of the movement to establish a northern branch of the Pharmaceutical Society in the city. RBGE had strong links with this branch, both in lecturing and in training pharmacists. In the 1860s, Baildon experimented with nature printing technology and demonstrated a new process using chromolithography. A local newspaper reported Baildon's achievement: "Mr. Baildon had effected a decided improvement on the method of nature-printing usually practised. The art of transferring impressions from plants to plates of lead or of softened copper, and from these producing prints in the natural colours of the plants, has been in use for a number of years … but none of these surpass, if they equal, in beauty or in accuracy the prints produced by Mr. Baildon's process." This underplays the work of Bradbury in London and Auer in Vienna, whose nature prints are clearer and more subtle than Baildon's. To give him his due, Baildon was using a different method and this in itself makes the work special, along with its Scottish origin and its scarcity. **GH**

British Algae: Cyanotype Impressions

The cyanotype photographic process was devised by Sir John Herschel and requires only water, art paper, two chemical salts and natural light. This made it an excellent medium for creating architectural blueprints. However, Anna Atkins (1799–1871), the daughter of the mineralogist, zoologist and chemist John George Children, used it to make images of seaweeds and ferns. Atkins' husband, John Petty Atkins, knew the pioneering British photographer William Henry Fox Talbot, but it is Anna Atkins herself who stakes a claim as the first person to produce a book with photographic illustrations. Over almost 10 years Atkins and her friend Anne Dixon, a cousin of Jane Austen, produced 12 albums of prints of British seaweeds. Each album contained nearly 400 cyanotypes. The work is an excellent marriage of science and art, and a monument to tenacity, dexterity and attention to detail, as each cyanotype needed to be exposed for anything between 10 and 40 minutes. Atkins sent out her work in parts, which eventually comprised three albums. The recipients were responsible for binding and collating the parts received. This perhaps explains why the set in the RBGE Library is complex and problematic. Volumes I & II contain 195 cyanotypes and were received as a duplicate from the Royal Botanic Gardens, Kew. The provenance of Volume III is unknown; it contains 179 cyanotypes and, rather curiously, a slip of lightweight blue paper inscribed: "A Mr W.J. Hooker et J.D. Hooker de la part de l'auteur". The question is, does it belong there? **GH**

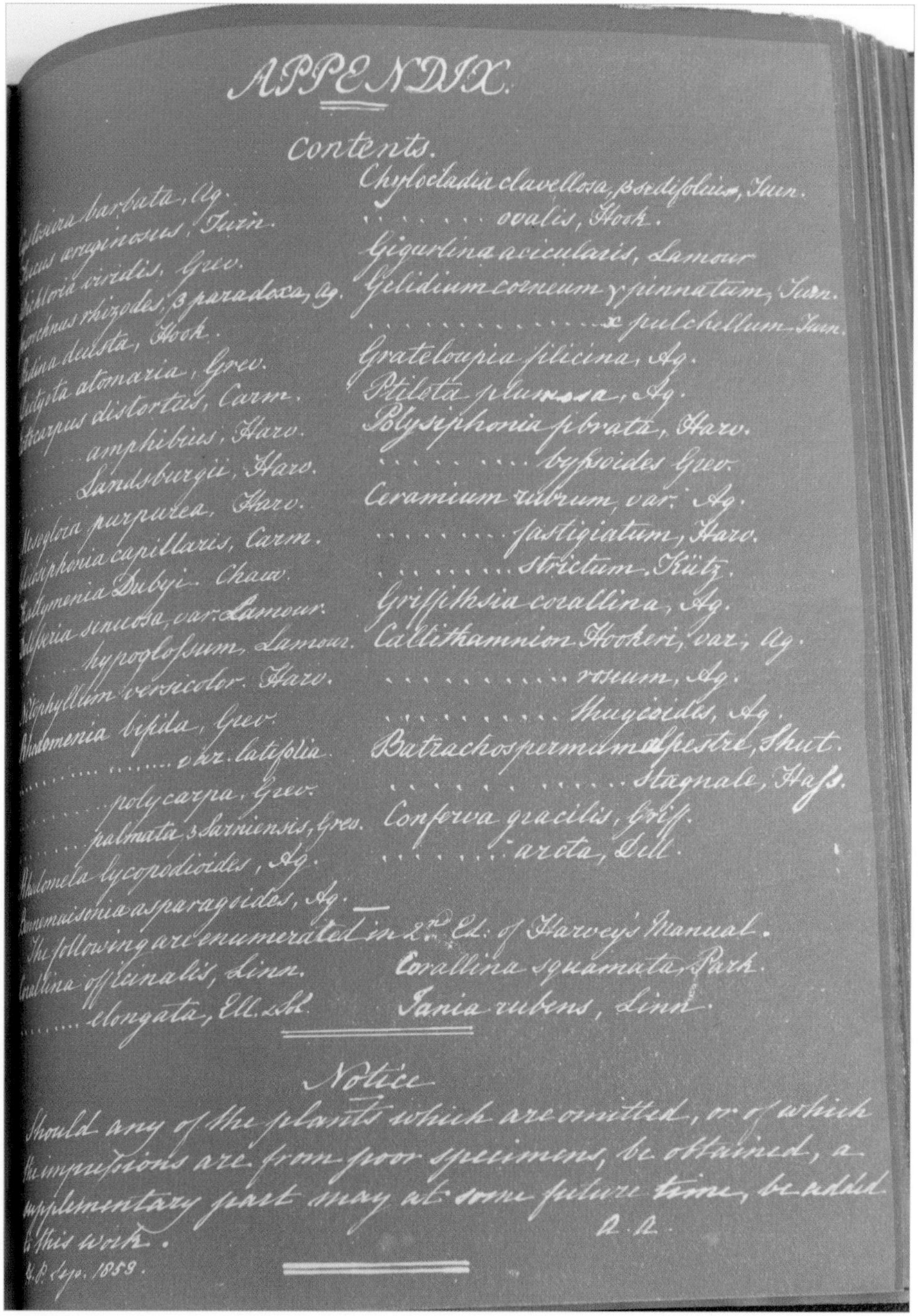
APPENDIX.

Contents.

barbata, Ag.
aeruginosus, Turn.
viridis, Grev.
rhizodes, β paradoxa, Ag.
deusta, Hook.
atomaria, Grev.
distortus, Carm.
amphibius, Harv.
Landsburgii, Harv.
purpurea, Harv.
capillaris, Carm.
Dubyi. Chauv.
sinuosa, var. Lamour.
hypoglossum, Lamour.
versicolor. Harv.
Rhodomenia bifida, Grev.
.... var. latifolia
.... polycarpa, Grev.
.... palmata, β Sarniensis, Grev.
Rhodomela lycopodioides, Ag.
Bonnemaisonia asparagoides, Ag.

Chylocladia clavellosa, Turn.
.... ovalis, Hook.
Gigartina acicularis, Lamour
Gelidium corneum γ pinnatum, Turn.
.... pulchellum Turn.
Grateloupia filicina, Ag.
Ptilota plumosa, Ag.
Polysiphonia fibrata, Harv.
.... byssoides Grev.
Ceramium rubrum, var. Ag.
.... fastigiatum, Harv.
.... strictum, Kütz.
Griffithsia corallina, Ag.
Callithamnion Hookeri, var. Ag.
.... roseum, Ag.
.... thuyoides, Ag.
Batrachospermum alpestre, Shutt.
.... stagnale, Hass.
Conferva gracilis, Griff.
.... arcta, Dill.

The following are enumerated in 2nd Ed. of Harvey's Manual.
Corallina officinalis, Linn. Corallina squamata, Park.
.... elongata, Ell. Sol. Jania rubens, Linn.

Notice

Should any of the plants which are omitted, or of which the impressions are from poor specimens, be obtained, a supplementary part may at some future time, be added to this work. A.A.

Green Olive Tree

Phillyrea latifolia

The idea of ornamenting herbarium sheets originated in the Netherlands in the early 1700s and may have been used to clearly identify the owner of a specimen to be loaned to friends and colleagues. The decorations were printed from copper plates and typically included a ribbon with the name of the species and a vase above which the specimen would be mounted. The example here is particularly ornate, with two cherubim holding the ribbon with the misspelled species name and an elaborate vase on a baroque stand with a monogram clearly visible. The identity of the original owner of this specimen is not yet known, although it is believed to have formed part of Paul Dietrich Giseke's herbarium. Decorated herbarium sheets had dropped out of fashion by the end of the 18th century and were replaced by more practical labels affixed to plain sheets. **EH**

Madagascar Laceleaf

Aponogeton madagascariensis

The beautiful fenestrated leaves of this aquatic species, commonly known as Madagascar laceleaf, explain why this plant is popular with aquarium enthusiasts around the world. Although it is labelled as collected by William Pool in Madagascar in 1876, it is possible that Pool's wife Mary was responsible for many of his collections, including this one. The species was first described in 1805, before Joseph Dalton Hooker transferred the species into the genus *Aponogeton* in 1883. It became very fashionable with the Victorians and its popularity continues today. **EH**

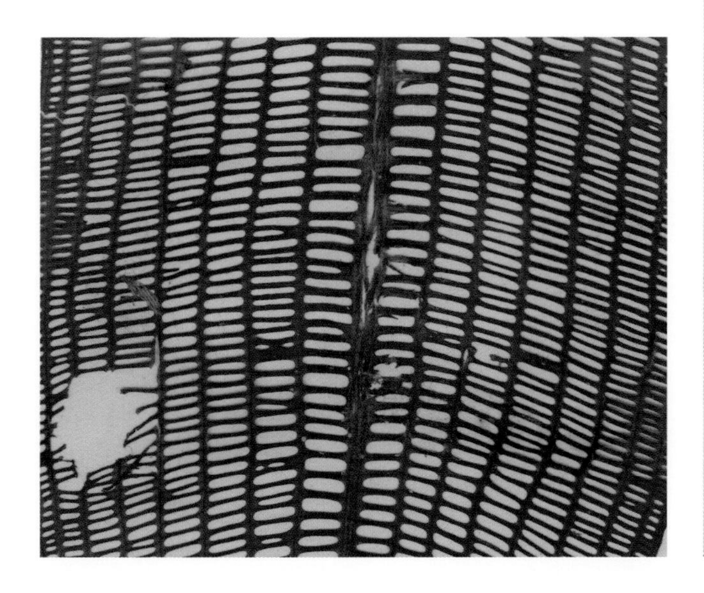

16
17
18
19
20

Scottish Biodiversity

Scotland has over 1,000 species of ferns, conifers and flowering plants, and its unique climate and landscape has resulted in a particularly rich flora of mosses, lichens, fungi and algae. This variety plays a critical role in Scottish biodiversity, creating a wide range of habitats, many of which are unique and under threat, such as the ancient forests and peat moorlands. The study of Scottish biodiversity has benefited from the work of RBGE botanists, from Robert Sibbald in the 17th century to those still pushing the boundaries of botanical research in the present day.

Scotia Illustrata

West Lothian landowner Robert Sibbald was the first Professor of Medicine at the University of Edinburgh and co-founder of the Royal College of Physicians Edinburgh and the Physic Garden which later became RBGE. Charles II appointed him as his physician and Geographer Royal for Scotland – the only person to be so honoured. In this capacity Sibbald planned a major geographic survey of Scotland, taking in landscape, economics, agriculture, natural history and ethnography. He gathered the information for his work by sending out pamphlets to selected people around Scotland, with a list of questions and directions. Written in Latin and published in 1684, *Scotia Illustrata* was the result of the survey. Amongst many other items it contains the earliest description of the Edinburgh Physic Garden on the Trinity Hospital site (now part of Edinburgh's Waverley Station). **GH**

Creeping Sibbaldia

Sibbaldia procumbens

Named by Linnaeus after Sir Robert Sibbald, this small, glaucous plant with tiny yellow petals grows at suitable altitudes around the northern hemisphere. In the mountains of Scotland it is found in areas where the winter snow lies for several months, protecting a distinctive plant community. Without snow, plants are exposed to severe frosts, searing winds and the risk of a mild spell in the spring which might stimulate premature growth only for it to be frosted later on. In some areas small plants with few leaves show signs of a changing climate and their health is a good indication of the winter snowfall. The main specimens on this sheet are very healthy and were collected in an area where snow-bed communities still occur. Fittingly, the earliest information on the plant in Scotland was recorded in Sibbald's *Scotia Illustrata* (p. 132); the plant is now used as RBGE's logo. **HM**

PINUS SYLVESTRIS, Linn.
Nat. Ord. Coniferae
Scots Pine

Scots Pine

Pinus sylvestris

Due to their size and form, conifers can often be very problematic to preserve as herbarium specimens. Here is an example of how effectively the Scots pine (*Pinus sylvestris*) can be preserved using an alcoholic solution. This is also a good example of how preserved specimens in this form can be used effectively in educational displays for students and the general public. *Pinus sylvestris* occurs from Western Europe across Asia to the Russian Far East, the most extensive distribution of any pine species in the world. Conifers have an important role at RBGE not just as living specimens but also in the Herbarium, where over 15,000 dried specimens provide the comparative material that is essential for studies in taxonomy, anatomy and conservation, as well as being used for teaching. In 2014, this iconic tree of the Scottish landscape was voted Scotland's national tree. MG

h feparate
›w only at
her large,
l covered
ne young
the pul-
›ranches
et been
3 inches
'ence at

Fig. 1. Fig. 2. Fig. 3. Fig. 4.

the outer than the inner fide, which refts next the ftem of the
t pyramidal apophyfis and a hooked umbo, the hook being
ed. Seeds about 4-10ths of an inch in length, dark brown

Pinetum Britannicum

In the 19th century, the Edinburgh-based nursery of Peter Lawson & Sons specialised in conifers and was one of the main commercial shareholders in two Edinburgh plant-collecting syndicates to North America: the Oregon Botanical Association and the British Columbia Botanical Association. Lawson's produced *Pinetum Britannicum: a descriptive account of hardy coniferous trees cultivated in Great Britain* as a way of showcasing new conifers to potential customers. The book was produced in a size known as elephant folio and each volume contains hand-coloured chromolithographs of botanical studies of cones and trees in native landscapes. A number of artists' drawings were used, including work by Robert Kaye Greville, James Black, William Richardson and Lady Canning.

The first 33 parts, funded by subscribers, were printed on the company's private press between 1863 and 1873, in a process managed by Edward James Ravenscroft (1816–1890), an Edinburgh-based printer and publisher. Publication was halted when Lawson's had to close and relaunch as a result of an ill-advised investment in fertilisers using the new miracle ingredient 'guano'. Ravenscroft took it upon himself to try to complete the planned work and a further 19 parts were issued between 1877 and 1884, jointly published in Edinburgh and London. **GH**

DEDICATED
BY COMMAND
TO THE
ILLUSTRIOUS MEMORY OF H.R.H. THE PRINCE CONSORT

PINETUM BRITANNICUM

A DESCRIPTIVE ACCOUNT OF ALL HARDY TREES OF THE PINE TRIBE CULTIVATED IN GREAT BRITAIN

WITH FAC-SIMILES OF THE ORIGINAL DRAWINGS MADE FOR THE WORK

Part XVI—PINUS CEMBRA

WITH ONE COLOURED PLATE, AND FOURTEEN ENGRAVINGS ON WOOD

PRIVATE PRESS OF PETER LAWSON & SON
EDINBURGH AND LONDON

PINUS BALFOURIANA.

FOX-TAIL PINE OF NEVADA.

IDENTIFICATION.—PINUS BALFOURIANA, *Oregon Committee's Circular* (1852); Gordon, *Pinetum*, 217 (1858); Engelmann in *Revision of Genus Pinus*, p. 19, and note 6, and in *Bot. of California*, ii. p. 125; *Gard. Chron.*, March 11, 1876, p. 332.
PINUS PARRYANA, Parlatore, cited by Gordon, *Pinetum*, ed. 2 (1875), p. 295.
PINUS QUADRIFOLIA, Parry, cited by Gordon, *Pinetum*, ed. 2 (1875), p. 295.

ENGRAVINGS.—*Cone, Seed, and Leaves*; *Gard. Chron.*, *loc. cit.*

Specific Character.—Pinus foliis quinis singulis falcatis brevibus; vaginis caducissimis; strobilis fascis elongatis attenuatis sub-pyriformibus parum curvatis, squamis sub-laxis, apophysi tetragono umbone transversim elliptico depresso; seminibus alatis, spermodermate maculato.

Habitat in California boreali.

A tree of about 80 feet high (seldom over 50, Engelmann). Branches said by Mr. Gordon to be pendulous and flexible (causa scientiæ non patet), bark smooth (deeply fissured, Engelmann), and reddish, leaves in fives (fig. 1), but varying upon the same shoot, there sometimes being only four, sometimes only three, and occasionally as few as two in the sheath. They are crowded on the branch, trigonal, short (about 1 inch long), ridged and slightly falcate, or curved inwards, without stomata on the back (fig. 2), with several rows on each of the inner sides (fig. 3). The margin is entire; sheaths caducous, composed of long scales; the hypoderm is double and the resin canals (fig. 4) peripheral; inflorescence not yet known; cones long (4 or 5 inches), tapering, and sometimes pear-shaped, but more blunt at the end (fig. 5), dark-brown, said by Mr. Gordon to be mostly solitary and pendent on the points of the branches and full of resinous matter; scales thin, flattened, with the apophysis tetragonal, depressed in the centre, in which is a transversely elliptical umbo. Seeds winged (fig. 6); wing large, straight at the back, and with a broad, bold curve on the other margin. Seed moderately large, projecting a little backwards from the wing; spermoderm maculated.

Fig. 5

Description.—We know nothing more of the appearance of this tree than the brief notices given by Jeffrey, that it grows to a height of 80 feet with a diameter of 3 feet. In his "Revision of the Genus Pinus" (1880), Engelmann states that this and *P. aristata*, in spite of the differences in the cones, are identical. In Utah and Nevada a form occurs with cones like those of *P. aristata*, but with short stout recurved prickles.

Geographical

[36] A

41

CEDRUS DEODARA,
A GROVE AT NACHAR NEAR THE WATERFALL.
From a Drawing by the late Lady Canning

Diatomacea
Auliscus
Cumbrae Sands
622

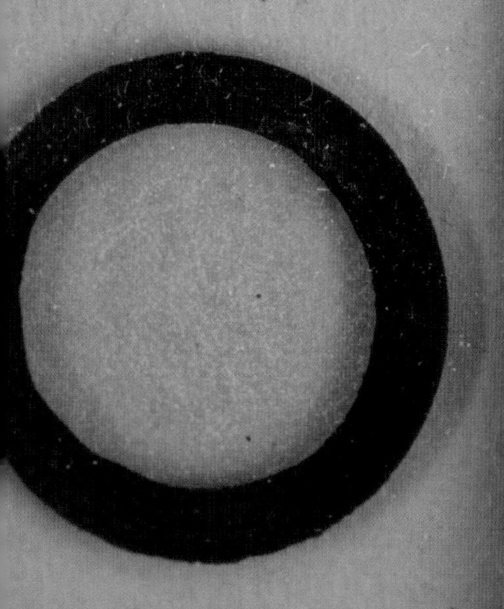

The Arnott Collection of Diatom Slides

The oxygen in every fifth breath we take is produced by diatoms. These microscopic organisms, with their unique cell walls made of silica, frequently form exquisite structures and are found in nearly every aquatic habitat. Given their importance to life on earth it is perhaps surprising how few people are responsible for our current knowledge of them. One of the most important figures in diatom research in the 19th century was George Arnott Walker Arnott. His passion for botany and his wide circle of botanical friends allowed him to develop one of the finest herbaria of the time, particularly rich in diatoms. In later life he devoted his energies to studying diatoms, and his slide collections, stored in beautifully crafted cabinets (p. 130), are now at RBGE. They are on permanent loan from the Scottish Association for Marine Science in Oban, who took over their custody from the Botany department of the University of Glasgow, where Arnott was Professor of Botany. **EH**

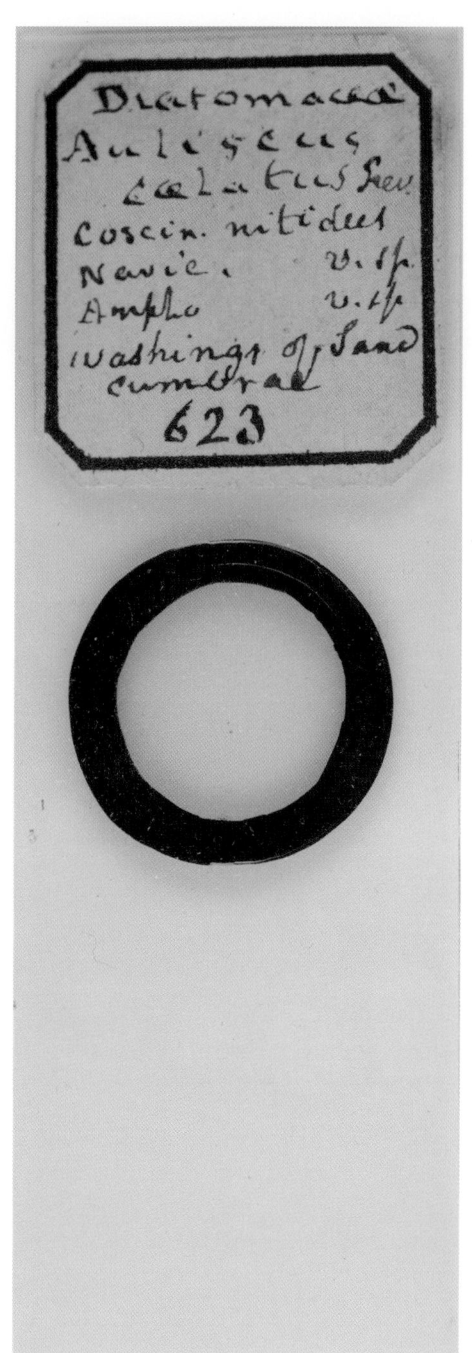

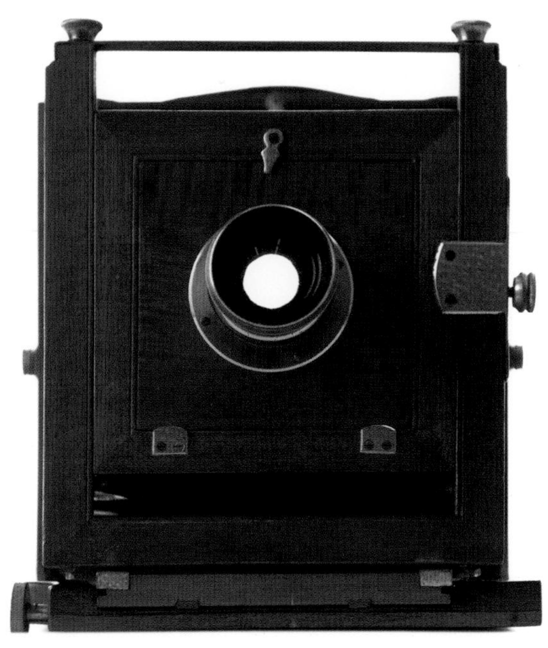

George Paxton's Camera

In 1894, Regius Keeper Isaac Bayley Balfour commissioned *Remarkable Trees in Ayrshire*, an album of the county's champion and heritage trees, from the Kilmarnock brewer and talented amateur photographer George Paxton (1850–1904). Paxton, who used his photographic skills to illustrate his journal articles on trees and provide lantern slides for lectures to local scientific societies, went on to become Photographic Artist to the Royal Scottish Arboricultural Society, recording the Society's summer excursions in Scotland, Ireland and Sweden. Over a century later, the gift by Paxton's grandson of his photographic collection, including his camera, has been a significant addition to the RBGE Archive. **HB**

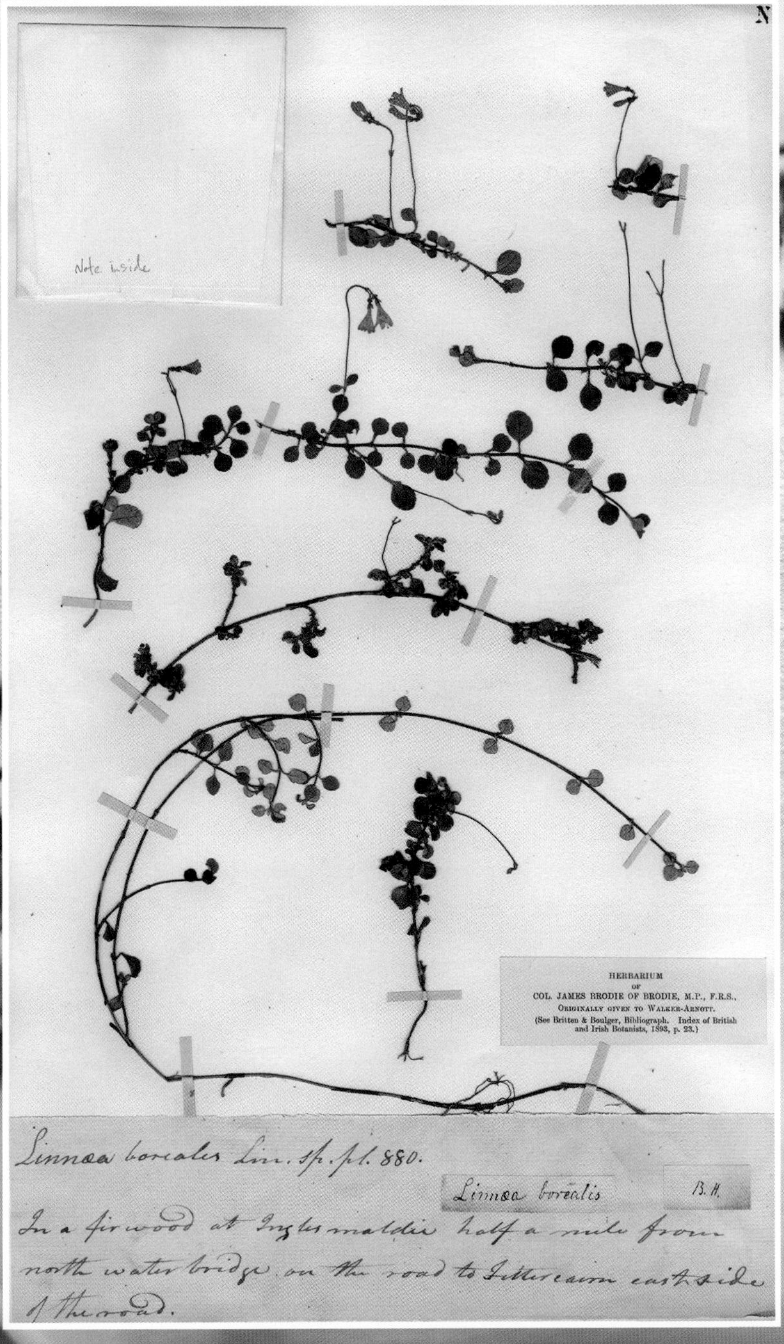

Twinflower

Linnaea borealis

In an uncharacteristic fit of (false) modesty Linnaeus chose this "little northern plant, long overlooked, depressed, abject, flowering early" to commemorate his own name. It was long overlooked in Scotland, being discovered neither by any of John Hope's students in their Scottish explorations, nor by the Rev. John Lightfoot in his 1772 tour that led to the first *Flora Scotica*. The twinflower was first discovered in Scotland by James Beattie, Professor of Civil and Natural History at Marischal College, Aberdeen, at Inglismaldie in Kincardineshire, in 1795. This specimen is very probably from that first collection, as Colonel James Brodie of Brodie was a friend of Beattie. Brodie's herbarium forms an important part of the RBGE Scottish collection and contains many early records including specimens from George Don (p. 142). Ironically, the Kincardineshire record probably represented an introduction – the plant is native to Caledonian pine forests, and its occurrence in eastern lowlands was probably a result of afforestation efforts by improving landlords. **HN**

77. Valeriana pyrenaica. Buxb. *cent.* ii. *t.* 11.

I firſt obſerved this plant, in 1782, by ditches and by the ſides of walls, near Blair-Adam, in Kinroſs-ſhire. I have alſo ſeen it in a wild ſtate near Glaſgow. I have ſince noticed it in one or two other places in Kinroſs-ſhire; and, ſome time ago (in company with Meſſrs Maughan and J. Neill) I found it in plenty, in a moiſt wood on the banks of the river Leith, about a mile below Collington, and three miles from Edinburgh; and in Sept. 1805, I obſerved it on the ſide of a rivulet, in a wood at Abercorn, Linlithgowſhire, (in company with Meſſrs P. Neill and Hoſey). It is remarkable of this plant, that it is much more luxuriant in its native than in a cultivated ſtate: hence I was obliged to ſelect the very ſmalleſt ſpecimens.—There can be no doubt whatever of this being a plant truly indigenous to Scotland.

Herbarium Britannicum

The fascicles of *Herbarium Britannicum* represent the major botanical legacy of George Don (1764–1814), and RBGE is fortunate to possess one of the handful of sets known to have survived intact. Each of the nine bound fascicles contains 25 dried specimens of flowering plants, mosses and lichens with printed labels giving habitat and locality information. Several of the specimens are first records for British plants. Also included are naturalised aliens, such as the umbellifer *Chaerophyllum aureum*. Don described three species new to science, for which the specimens represent the original voucher or 'type' specimens. Although in the tradition of the bound *Hortus Siccus* (p. 55), because it has a printed title page the work is also considered to be a book and is a fascinating hybrid artefact of the late Scottish Enlightenment. Several of the parts were issued while Don was Principal Gardener at RBGE on its Leith Walk site, and were almost certainly compiled in the gardener's cottage where he lived with his family. Two of his sons gained international botanical renown; George Don junior (1798–1856) was one of the first professional plant collectors for the [Royal] Horticultural Society and a botanical author, while David Don (1799–1841) wrote *Prodromus Florae Nepalensis* in 1825 and became the first Professor of Botany at King's College London. **AE**

77. Valeriana pyrenaica. Buxb. *cent.* ii. *t.* 11.

I firft obferved this plant, in 1781, by ditches and by the fides of walls, near Blair-Adam, in Kinrofs-fhire. I have alfo feen it in a wild ftate near Glafgow. I have fince noticed it in one or two other places in Kinrofs-fhire; and, fome time ago (in company with Meffrs Maughan and J. Neill) I found it in plenty, in a moift wood on the banks of the river Leith, about a mile below Collington, and three miles from Edinburgh; and in Sept. 1805, I obferved it on the fide of a rivulet, in a wood at Abercorn, Linlithgowfhire, (in company with Meffrs P. Neill and Hofey). It is remarkable of this plant, that it is much more luxuriant in its native than in a cultivated ftate: hence I was obliged to felect the very fmalleft fpecimens.—There can be no doubt whatever of this being a plant truly indigenous to Scotland.

Scottish Primrose

Primula scotica

As most of mainland Scotland was covered by glaciers we have few endemic species that are found nowhere else in the world. This special little primula is now restricted to the north coast of Scotland and on Orkney, where it grows on the short grassland of cliff edges and grazed areas near the coast. The number of specimens on this sheet shows how abundant it has been but illustrates the threat from botanists in the past who collected to excess. Fortunately many populations have survived but they depend on sensitive grazing management of base-rich areas and a suitable climate. *Primula scotica* flowers best after mild winters and during warm summers. The increasing number of windy days associated with climate change is of concern, as this will inhibit pollinators and reduce seed set. Without regular recruitment from seed, small populations can be lost and larger populations gradually dwindle. **HM**

N. B. R.
Professor Balfour's
PUPIL.
20th July 1872.
EDINBURGH
TO
DUNKELD
Via FIFE.
THIRD CLASS.
8
2
VICTORIA
STEAMER.
and Alloa to Granton Pier.
One Shilling and Sixpence

John Hutton Balfour's Excursion Notebooks

In 1845, John Hutton Balfour was given the appointment of Professor of Botany at the University of Edinburgh, and then the Regius Keepership of RBGE, mainly due to his reputation as a thorough and innovative teacher. He used teaching diagrams and models (p. 28) as well as microscopy, and, following in the footsteps of his own teacher, former Regius Keeper Robert Graham, he took his pupils on numerous plant-hunting excursions, often utilising the new train networks to do so. His notebooks from these excursions give some insight into the challenges of 'botanising' in Scotland and beyond in the Victorian period. **LP**

Professor Balfour & Party

Royal

Dalwhinnie

W. PULLAR

1867

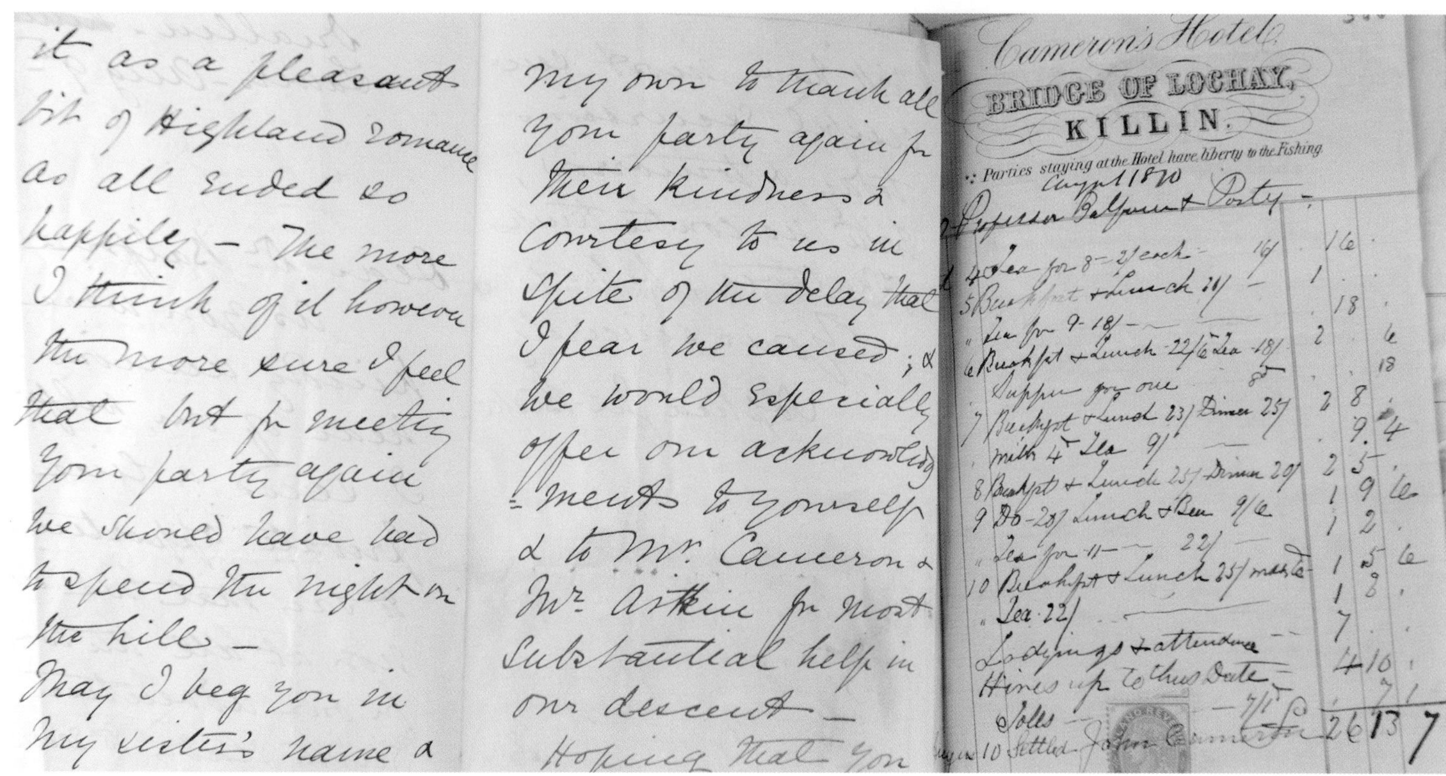
it as a pleasant bit of Highland romance as all ended so happily – The more I think of it however the more sure I feel that but for meeting your party again we should have had to spend the night on the hill –
May I beg you in my sister's name & my own to thank all your party again for their kindness & courtesy to us in spite of the delay that I fear we caused; & we would especially offer our acknowledgments to yourself & to Mr Cameron & Mr Aitkin for most substantial help in our descent –
Hoping that you

Cameron's Hotel,
BRIDGE OF LOCHAY,
KILLIN.
Parties staying at the Hotel have liberty to the Fishing

Professor Balfour & Party

Lodgings & attendance

Hires up to this Date

Robert Kaye Greville's Scottish Landscape Drawings

The biodiversity demonstrated in this section owes much to the varied Scottish landscape, captured beautifully here in drawings from a collection by Robert Kaye Greville. The 63 graphite drawings, created between 1834 and 1852, are on toned paper with white gouache highlights and consist mainly of landscapes from Scotland, with some views of Wales and the north of England. The drawings shown here include views of Edinburgh, Sutherland and the Invergarry Bridge over the Caledonian Canal. The image of Sutherland is dated 1834 when Greville travelled in the area with John Jardine, Sir William Jardine, Prideaux John Selby and James Wilson. The collection was donated to the RBGE Library in 1920 by Greville's family and a selection was exhibited at a meeting of the Botanical Society of Edinburgh by Mr William Evans on 18 November 1920. **LM**

Invergarry Bridge
Caledonian Canal. Sept. 1848
RKL

Dead Man's Rope

Chorda filum

Also known as 'dead man's rope', *Chorda filum* grows in the sheltered rock pools of the British shoreline. How do you mount a specimen of seaweed which grows up to 8 metres long on a piece of card 20 centimetres long? This specimen was collected by Dr Drummond in Belfast and shows the care that many collectors take over the mounting of their collections. Some of the most beautifully mounted specimens are to be found in the Greville herbarium. Robert Kaye Greville's significant herbarium was bought by the University of Edinburgh and subsequently formed a founding collection of the RBGE Herbarium. **EH**

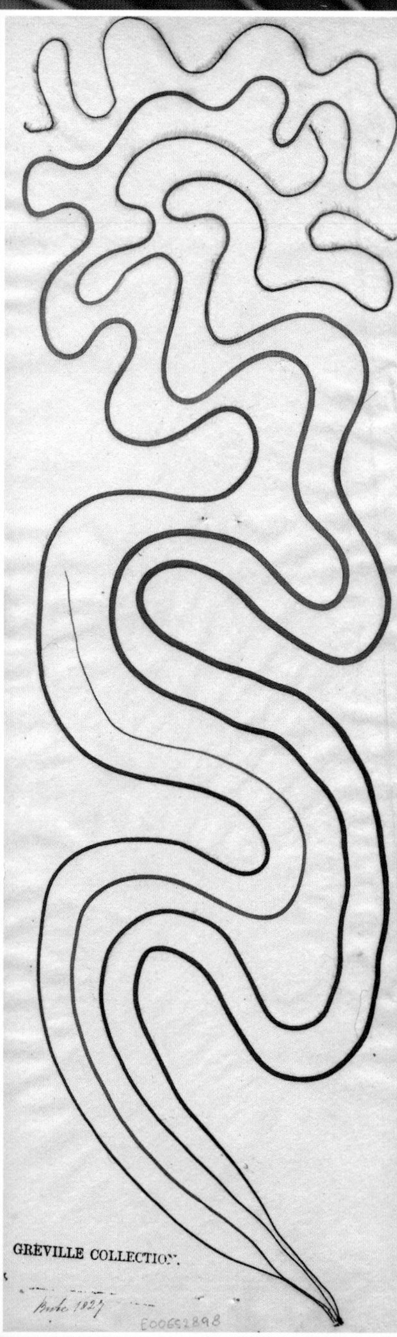

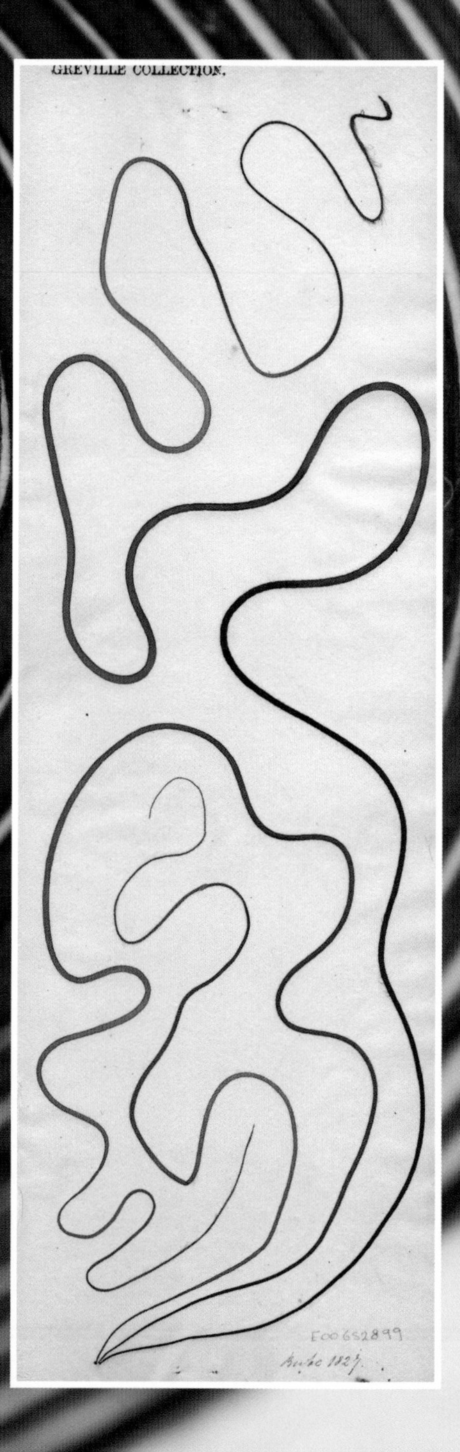

Magellanic Bog-moss

Sphagnum magellanicum

A widespread moss found in the bogs of Scotland was perhaps an unusual plant to find being exported in vast quantities to the battlefields and hospitals of the First World War. However, its absorbent and antiseptic properties, as well as its ready availability, resulted in nearly a million sphagnum dressings being sent each month from Britain for the wounded, a process illustrated here by photographs from RBGE's Archive. Over 2,000 years earlier, a similar dressing was pressed to the chest of a Bronze Age warrior, whose burial site was later discovered in Fife, and these dressings were extensively used during the battle of Flodden in 1513. This specimen was collected in the lowlands of Scotland by James McAndrew (1836–1917) who did not survive to see the end of the war. **EH**

Biodiversity at Risk

Species become extinct for many reasons, over-collecting, habitat loss and climate change among them. In some cases the only record of a species ever having been alive may be a specimen preserved in a herbarium somewhere in the world. The continuing work of botanists and taxonomists is to discover the plants which exist before we lose them, but the greater aim is always to prevent their loss. The collections held at RBGE provide information which not only helps record the world's biodiversity but also conserves it for the future.

Cotton Deer Grass

Trichophorum alpinum

The only certain locality for this plant as a British native was the Moss of Restenneth, near Forfar, where it was discovered by George Don in July 1791, and shown by him to a fellow Angus botanist, the great Robert Brown, later the same year. Many specimens were collected which circulated in herbaria for decades, but Don himself recorded that the plant was extinct by 1804, as a result of draining the bog to extract marl (for use as agricultural lime) and the cutting of peat (for fuel). The plant has a wide distribution in northern Europe and North America, and is common on raised bogs in Scandinavia – in Scotland it must have been on the edge of its range and hence rare. There have been other possible Scottish records, such as a collection apparently made by John Hutton Balfour at Durness in Sutherland in 1827, but which may have been due to a mix-up in the herbarium. However, as the habitat is widespread in Scotland, and the plant rather inconspicuous (even when in fruit, when the hairs surrounding the achenes lengthen), it may still occur and await rediscovery. **HN**

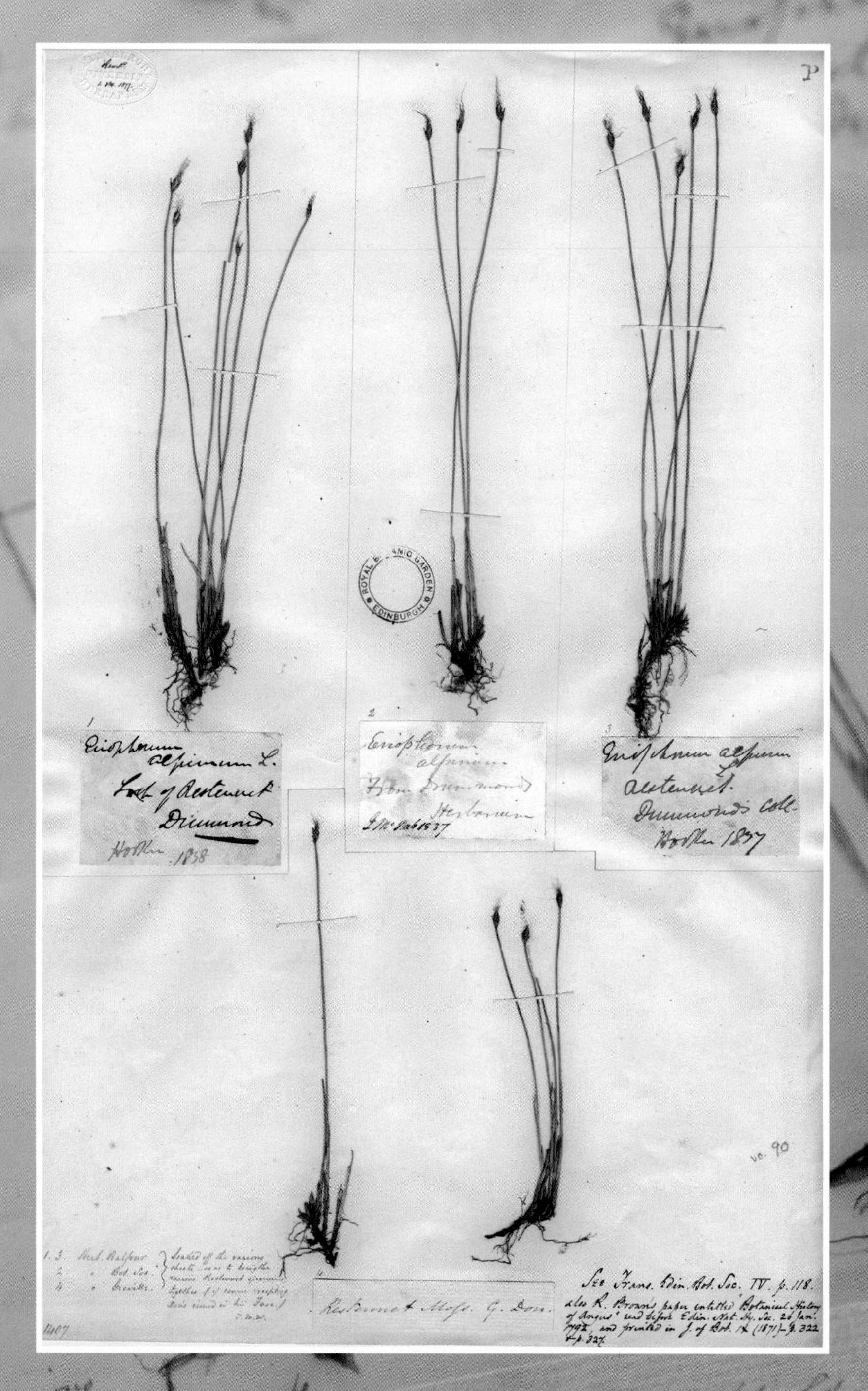

Alpine Butterwort

Pinguicula alpina

Like *Trichophorum alpinum*, this is a widespread north-temperate plant which became extinct in the 19th century at what was its single British locality. The distribution of this white-flowered butterwort is from Europe, through the Himalaya, to western China – so in Scotland it was at the edge of its range. It was discovered in 1831 by the Rev. George Gordon on the estate of Rosehaugh on the Black Isle, Easter Ross. That it was collected in large numbers can be seen from this sheet (one of 17 in the RBGE Herbarium, bearing between them no fewer than 117 individual plants), and it was apparently found at several sites in the vicinity. The rarity was eventually confined to a single boggy patch at Avoch, around which the landowner erected a protective wall. The Oxford botanist George Claridge Druce saw it there in 1882, but by 1919 was able to pronounce its extinction. The reasons given were drainage for conversion of the original habitat into arable land and the drying out of the final bog by the invasion of conifer seedlings. **HN**

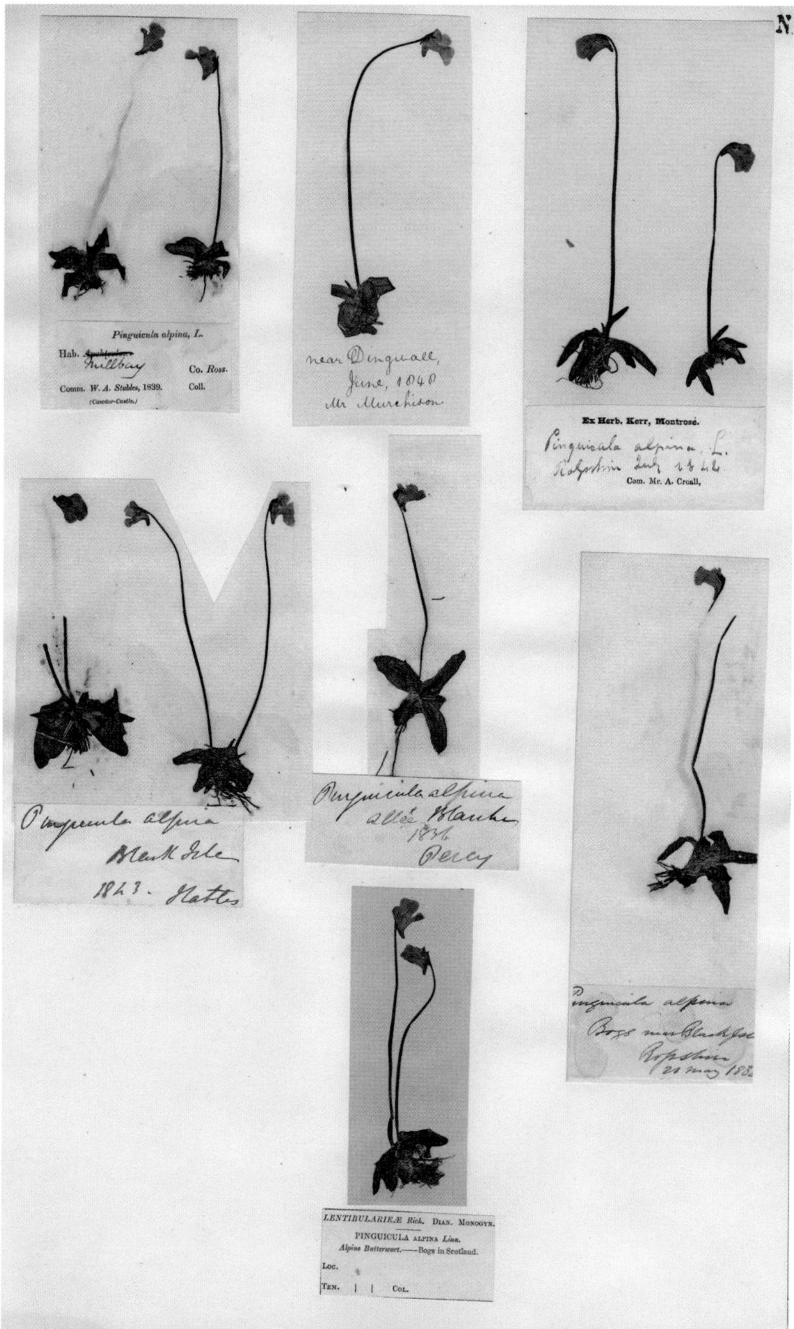

Begonia bracteata

This initially unprepossessing *Begonia* grows only at the base of Gunung Bungkuk, an isolated basalt plug in Bengkulu Province, southern Sumatra. The specimen here represents the only collection known to science. The species was first collected by William Jack (see p. 105) in June 1821, and described in his *Malayan Miscellanies* in 1822. The species became the type of the eponymous *Begonia* section *Bracteibegonia*, which potentially still contains hundreds of undescribed species. Jack wrote that during the ascent of Gunung Bungkuk his guides feared "the vengeance of the evil spirits if they conducted strangers to the summit; they were, therefore, advising to return at every difficulty, and the ascent was ultimately accomplished without their aid, or rather in spite of them". Within a year of the ascent Jack was dead and his precious herbarium destroyed by fire aboard the *Fame*. Our specimen was collected by a member of RBGE staff in collaboration with Herbarium Bogoriense, during a 2010 expedition following in Jack's footsteps, and is the world's first encounter with this rare and enigmatic plant for nearly two centuries. **MH**

HERB. HORT. EDINB.
NEOTYPE
of Begonia bracteata Jack
selected by M. Hughes & D. Girmansyah
in Gard. Bull. Sing. 63: 83-96 (2011)
ROYAL BOTANIC GARDEN EDINBURGH
E00416880
Flora of Sumatra
Royal Botanic Garden Edinburgh and Herbarium Bogoriense
Begoniaceae
Begonia bracteata Jack
INDONESIA: Bengkulu: Gunung Bungkuk. 610 m. 3°35'3.12"S, 102° 25'24.96"E. Sporadically common forest floor herb on very steep mountainside.
Male flower with 4 tepals, white; female with 5, white turning green with reddish blush. Fruit recurved when mature, placentae bifid. .
D. Girmansyah & M. Hughes No. DEDEN1495
15 Aug 2010

St Helena Olive

Nesiota elliptica

Nesiota elliptica, the St Helena olive, is the only species in its genus and was one of only 51 flowering plant species endemic to the South Atlantic island of St Helena. This island was the subject of pioneering studies on island conservation by Professor Quentin Cronk, formerly of RBGE. In the middle of the 19th century, only 12 individuals were known in the wild and over the next 150 years the numbers declined, with the last remaining individual dying in 1994. The species, and thus the genus, became extinct in 2003 when cultivated individuals were lost as a result of fungal infections. While studying for a PhD at Kew, RBGE's James Richardson used DNA sequence data to determine that *Nesiota* had been an ancient and isolated lineage (related to genera from South Africa) whose loss was thus considered even more significant. **JR**

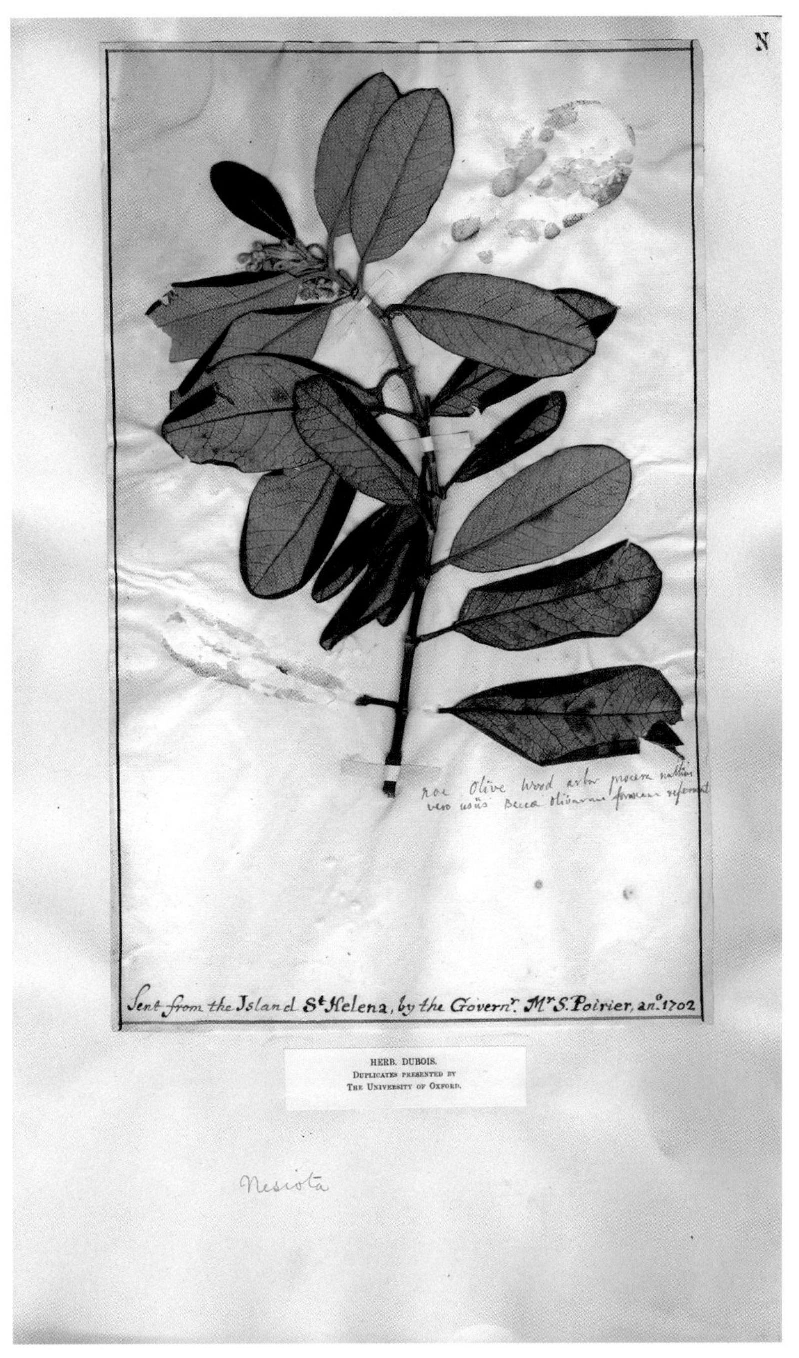

Fifteenth-century Wattles

Why are these dusty old sticks considered treasures? These so-called wattles are sticks of hazel that were cut and bent to make a wattle-and-daub wall panel on the site of a building erected more than 500 years ago. That means that the lichens that make the lovely patterns on the bark are historic too, offering a unique insight into historic biodiversity and how environmental conditions have changed since that time. Lichens are fantastically sensitive to their environments and very well studied, particularly in Britain. Scientists at RBGE use them to study changes in biodiversity. This set of wattles was collected from inside a building erected in the 15th century in Wiltshire and, along with more than 150 other pre-industrial collections of epiphytes made by archaeological examination of historic buildings, tells a fascinating tale of climate change and biodiversity loss attributable to the effects of the Industrial Revolution. Although the tropics are at the forefront of current concerns about biodiversity loss, if you turned the clocks back a couple of centuries, you would see that it happened here in the temperate zone too. **RY**

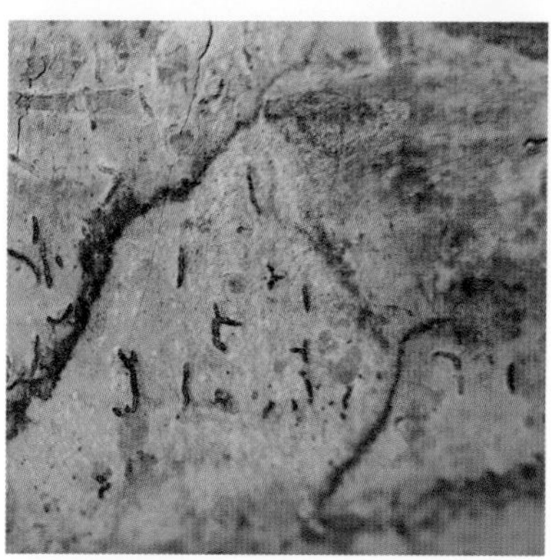

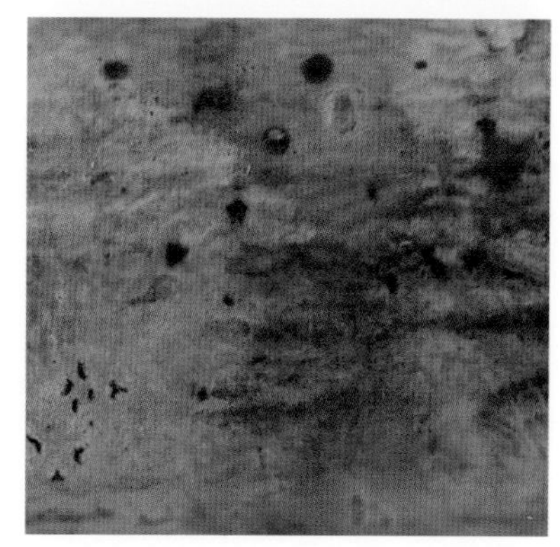

Acknowledgements

The Royal Botanic Garden Edinburgh would like to thank the Andrew W. Mellon Foundation for their support for the RBGE Herbarium's digitisation programme over the last 10 years, and the Sibbald Trust for their support for the associated Botanical Treasures exhibition at the John Hope Gateway, 1 July–30 September 2014.

We would like to thank Alice Young for her work on the editing, Caroline Muir for the design and Amy Fokinther for the photography as well as all of the staff and contributors who brought their knowledge and enthusiasm to the project.

Further Information

The following resources may be useful for anyone who would like further information on the Royal Botanic Garden Edinburgh and the collections held in the Herbarium and Library:

Bown, D., *4 Gardens in One: the Royal Botanic Garden Edinburgh* (HMSO, Edinburgh, 1992)
Fletcher, H.R. & Brown, W.H., *The Royal Botanic Garden Edinburgh 1670–1970* (HMSO, Edinburgh, 1970)
Mathew, M.V., *The History of the Royal Botanic Garden Library Edinburgh* (HMSO, Edinburgh, 1987)
Rae, D., *The Living Collection* (Royal Botanic Garden Edinburgh, Edinburgh, 2011)

The RBGE website has links to a number of resources that may be of interest, including:

The Herbarium Catalogue – available at
http://www.rbge.org.uk/databases/herbarium-catalogue

The Library Catalogue – available at
http://www.rbge.org.uk/databases/library-catalogue

Botanics Stories (stories and updates on RBGE collections, events and research) – available at
http://stories.rbge.org.uk/

The Library and Herbarium are open to bona fide researchers and information on visiting them can be found at:

Herbarium:
http://www.rbge.org.uk/science/herbarium/visitor-information

Library & Archives:
http://www.rbge.org.uk/science/library-and-archives/using-the-library